An introduction to engineering materials

An introduction to engineering materials

L.M. Gourd
B.Sc.(Eng.), C.Eng., M.I.M., F. Weld. I, M.I.T.D.
The Welding Institute

Edward Arnold

First published 1982
by Edward Arnold (Publishers) Ltd
41 Bedford Square, London WC1B 3DQ

ISBN 0 7131 3444 5

British Library Cataloguing in Publication Data

Gourd, L.M.
An introduction to engineering materials.
1. Materials
I. Title
620.1′1 TA403

ISBN 0-7131-3444-5

Filmset in 10/11pt English Times 501 by Colset Private Ltd and printed in Great Britain by Richard Clay (The Chaucer Press) Ltd, Bungay, Suffolk.

Contents

Preface

Anyone who writes a book for engineers about materials runs the risk of providing critics with a field-day – there are as many views on what it should contain as there are teachers of the subject. Unless the book slavishly follows a published syllabus, it will inevitably reflect the author's background. In this book, I have presented a very personal view of materials based on my own experience as a metallurgist in the engineering and manufacturing industries. I have tried to tell the story of engineering materials. No doubt, in our present world of high-sounding educational jargon, this is an outmoded concept; but, for me, the way in which we can make materials answer our needs is still a fascinating story. If it does nothing else, I hope that this book will arouse readers' interest and infect them with some of my enthusiasm for the subject.

In planning the text, I have started from the simple premise that engineers only want to know about materials because they use them. They are unlikely to find any value in an academic study of materials science in isolation, but they are undoubtedly interested in properties and in the practicalities of how they can be controlled. I have attempted, therefore, not only to lay the foundations for an understanding of structure–property relationships in commonly used metals and plastics, but also to relate these concepts to readily identifiable applications. Similarly, in dealing with forming and casting processes I have introduced only those principles of operation which are needed to show how material properties influence the choice of manufacturing techniques and how these in turn determine the properties of the finished product.

Many people may be surprised to see the inclusion of equilibrium diagrams. To a large extent this reflects the requirements of a number of syllabuses. Used sensibly, however, equilibrium diagrams can help the student to recognise the factors influencing structure and hence properties. The danger is that their study can quickly become an end in itself with the result that their limitations become obscured. With this in mind, I have been careful to select only those parts of typical diagrams which illustrate specific points, and I have introduced the concept of non-equilibrium to prepare the reader for further study of real-life engineering situations.

This book is intended to provide a self-contained, introductory course, and I have assumed only an elementary knowledge of physical science as a prerequisite. The principal aim is to cover the syllabuses for level-2 units on materials in Technician Education Council and Scottish Technical Education Council certificate and diploma programmes. In particular, it deals with all the objectives in the TEC standard unit Materials technology II (U80/738). I hope, however, that students on other courses will find that this book provides a useful introduction to the study of engineering materials.

I must acknowledge my indebtedness to Frank Toplis (chairman of the TEC A6 programme committee) and John McCormick (Welding Institute) who kindly read the text and offered many very helpful comments; to my wife, for her patience in translating my scribblings into a readable typescript; and to Bob Davenport, of Edward Arnold (Publishers) Ltd, for his invaluable help in editing the manuscript.

L.M. Gourd

Acknowledgements

The author and publishers would like to thank the following for their kind permission to reproduce photographs or illustrations:

British Rail and Oxford Publishing Co. Ltd (fig. 1.2), The Mansell Collection Ltd (fig. 1.3), Mr John Reynolds and Hugh Evelyn Ltd (fig. 1.4, from *Windmills and watermills*), Ironbridge Gorge Museum Trust (fig. 1.5), the Welding Institute (figs 3.9, 9.5, 10.1, 10.3, and 10.11), Avery-Denison Ltd (figs 4.2 and 4.14), Vickers Instruments Ltd (fig. 7.9), Aical Ltd and P.I. Castings (Altrincham) Ltd (fig. 8.3), the Zinc Development Association (fig. 8.5), the Copper Development Association (fig. 8.9), Delta Extruded Metals Co. Ltd (fig. 9.2), and Professor R.W.K. Honeycombe and R.A. Ricks (fig. 10.6).

The following are reproduced from or based on BSI publications by kind permission of the British Standards Institution, 2 Park Street, London W1A 2BS, from whom copies of the complete publications may be obtained:

Table 4.1 (*BSI Yearbook*), fig. 4.3 (BS 18:part 2:1971), Table 8.1 (BS 1490:1970), Table 8.2 (BS 1004:1972), Table 8.4 (DD 38:1974), Tables 9.3 and 9.4 (BS 970:part 1:1972), Table 9.5 (BS 4360:1979), Table 9.7 (BS 1470:1972), Table 10.4 (BS 4360:1979), Table 10.5 (BS 3100:1976), and Table 10.6 (BS 970:part 1:1972).

1 Materials in engineering

1.1 Introduction

For most of man's history there have been only a few useful constructional materials. Stone and wood have featured prominently throughout the ages. Metals like bronze and iron have also played a very important role, but until the middle of the eighteenth century they were in limited supply and were therefore relatively expensive. In the main, metals could be used only for small components, due to the limitations of manufacturing techniques – guns and church bells were probably the largest items regularly made from metals. Man had to make the most of the properties these materials offered, which meant that designs were dictated or perhaps limited by the materials available.

This interaction between available materials and design can readily be seen in the development of buildings over the ages. From the earliest primitive mud huts to the eighteenth-century timbered cottages of Suffolk, we can observe the builders' dependence on local supplies. Granite or sandstone blocks were in common use in upland and mountainous regions, while in other areas the principal construction material was flint which could be dug from chalk hills. The art of brickmaking was also developed where there were deposits of suitable clays, and in wooded areas timber-framed buildings with a lath-and-mortar cladding were the norm. Similarly, slate could be mined in some mountains and was used for roofs, but in the marshy plains it was found only in the houses of those people who could afford the high cost of transport. Poorer folk thatched the roofs of their houses with reeds collected from nearby marshland.

Bridges have always played an important part in the development of links between communities and the spread of ideas. Right up to the nineteenth century they also show the controlling role of the availability of materials. The problem which has always confronted the bridge builder is the need to span wide rivers and gorges. The earliest and easiest solution was to fell a tree across the obstacle – always supposing a tree was available in the right place. The success of this technique depended on the types of tree growing in the locality. Once the idea of a keyed arch had been developed (fig. 1.1), the way lay open to build bridges of stone and brick. This concept was of major importance. It enabled the construction of not only strong bridges to take place, but also of buildings with wide roofs. Probably one of the most spectacular achievements in the use of an arched structure is the great dome of the basilica of St Sophia at Constantinople (Istanbul). This was built by the Roman emperor Justinian in 530 A.D. and has a diameter of 32.6 m (fig. 1.3). Thirteen hundred years

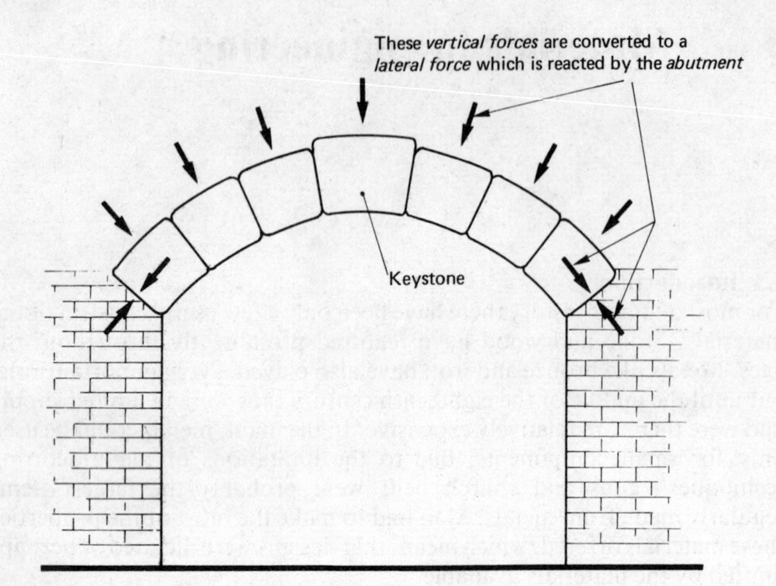

Fig. 1.1 Principle of a keyed arch

Fig. 1.2 Brick bridge built by Brunel at Maidenhead

Fig. 1.3 Inside of the dome of the basilica of St Sophia

later, Brunel was using the same basic principles to build a bridge with the largest brickwork span in the world (39 m) over the River Thames at Maidenhead (fig. 1.2).

The enforced reliance on available materials led to the development of many ingenious ideas, not only in buildings and bridges but also in many other areas. Naval architects of the seventeenth, eighteenth, and early nineteenth centuries showed a sophisticated understanding of the use of wood as a structural material. Wood was an almost universally available material and was also used in mechanical systems. It was employed to great effect in making large cog wheels for windmills (fig. 1.4).

Fig. 1.4 Wooden brake wheel and lantern pinion in a Dutch mill. (Note how the spokes and teeth are cut so that the force is applied at right angles to the grain, to get maximum strength.)

Much of the approach to design in these periods was based on knowledge gained by experience. It is difficult for us completely to appreciate how that experience was gained. Nowadays, with modern systems of communication, the news of a bridge collapsing is flashed around the world in a matter of minutes. We have no record of how many stone bridges fell down before the knowledge needed to construct a keyed arch was gained by the early builders. It is fairly safe to suggest that, although a few far-sighted men attempted to make an analytical approach, most designs were developed by trial and error. This made it difficult for engineers to adapt to new materials.

The degree of dependence on well tried and trusted concepts is shown in the Iron Bridge which was built at Coalbrookedale, Shropshire, in 1779 (fig. 1.5). Improvements in the methods used to produce iron and to cast it

Fig. 1.5 The Iron Bridge in Shropshire

into useful shapes offered the bridge builder a new material with properties very different from stone, brick, or wood. In the event, the Iron Bridge was built to the same design as that developed over the years for brick structures. It took time for engineers to adapt their thinking to make the best use of the new materials which came with the industrial revolution.

Most of the history books concentrate on the growth in the use of machinery during the industrial revolution and on the sociological effects of changing patterns of work and domestic consumption of goods which could be produced more cheaply. It is easy to overlook the significant changes that took place in the approach to design. The great contribution made by men like Brunel, Telford, and Stephenson was not so much that they changed the whole concept of communications but that they challenged many traditional concepts. Along with other nineteenth-century engineers, they gradually moved away from the practice of asking how they could use a material to do what was wanted – instead, they started to specify the properties they needed in a material to achieve their objectives. At the same time, the transport systems they built enabled materials to be moved easily from one part of the country to another, and regional differences in materials usage started to disappear.

The impetus this gave to the development of new materials has benefited today's engineers. Indeed, it may well be that some future historian looking back on the period from 1779 to the present time will consider that our greatest achievements were not controlling nuclear power and travelling into space but rather the ability to produce the materials on which those feats depended. Metals which stand up to the environment in a nuclear reactor, heat-resistant tiles to enable the space shuttle to re-enter the earth's atmosphere without burning up, and semiconductors for sophisticated computer circuits are examples of materials which provide the key to viable designs. Without them, some of our greatest achievements would not have been possible.

The modern engineer has a vast range of materials at his disposal: strong metals which can be fabricated; plastics which resist attack by acids and can be produced in a variety of shapes; ceramics which withstand very high temperatures. Effective design in engineering requires that we should be able to put these materials to the best use. We must choose the right material for the job. To do this, we need to know how and why different materials behave differently in service. Metallurgists, polymer scientists, and other specialists can give guidance, but the engineer must understand the principles on which material behaviour is based if this advice is to be of value.

In this book we will explore the structure of common engineering materials and see how this controls their properties. Typical materials will also be considered to illustrate the variations in properties that can be achieved by careful selection and treatment. Before this can be done, however, we need to establish the factors which influence the selection of a suitable material.

1.2 Selecting materials

Broadly speaking, materials used in engineering fall into three groups:

a) metals;
b) organic materials, e.g. plastics and wood;
c) ceramics, e.g. glass, bricks, and concrete.

A component or fabrication can be made entirely from a single material within one of these groups. Frequently, however, more than one material may be used to achieve the optimum blend of properties.

The choice of materials involves consideration of a number of factors which can be grouped under the following headings:

i) service requirements,
ii) manufacturing requirements and cost,
iii) cost of raw or bulk materials.

While this division is useful for our purposes, in practice the different aspects cannot be separated. They interact with each other, and there is rarely a unique suitable choice of material. More often than not there are two or three possibilities, and the final decision may be determined by the preference and experience of the designer.

Service requirements

Probably the first question which has to be asked is, 'What must the component do in service?' The answer to this gives the clue to the properties which a suitable material must have. Armed with this information, the materials specialist can make recommendations or the materials manufacturer can be asked to suggest a suitable product.

Strength In most applications, some consideration must be given to the strength of the material. As an example, we can consider a tubular component like a cylinder to hold compressed gas which is subjected to high internal pressures, say 150 bar (fig. 1.6). A strong metal must be used, so that the cylinder will not burst open. On the other hand, a cylinder may simply be a water container, in which case the only strength needed is that required to prevent the walls collapsing under the static pressure of the water.

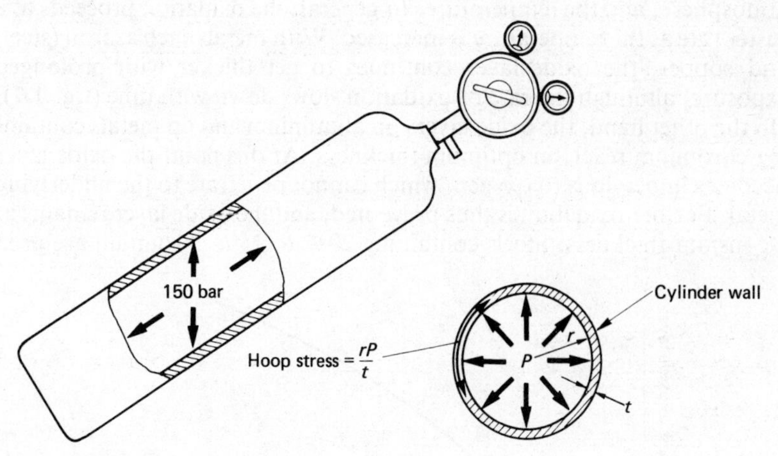

Fig. 1.6 Pressure inside a gas cylinder

The manner in which a load is applied profoundly affects the way in which we specify strength requirements. A cantilever beam behaves differently when subjected to a force which is steady or is applied suddenly or fluctuates.

Strength is a very important consideration in choosing a material, and we will be studying its various aspects in more detail in chapters 3 and 4.

Corrosion resistance Many service environments are hostile to materials and chemically attack the surface. This attack takes a variety of forms which are grouped under the general title 'corrosion'.

In some cases the material is dissolved by a liquid with which it is in contact. We cannot make storage vessels for nitric acid from carbon steel because the acid reacts with this metal, thus reducing its thickness. Instead, a stainless steel, containing chromium and nickel, or a plastics material must be used. Some plastics are soluble in alcohol and must not be included in systems handling this liquid. Data on the resistance of materials to attack by various chemicals is readily available, and selection on this basis is relatively straightforward.

Oxidation is a more specific form of chemical attack. Metals exposed to air or oxidising gases may react at the surface with oxygen,

e.g. iron + oxygen ⟶ iron oxide

$2Fe + O_2 \longrightarrow 2FeO$

aluminium + oxygen ⟶ aluminium oxide

$4Al + 3O_2 \longrightarrow 2Al_2O_3$

The oxide so produced grows at a rate which depends on the metal, the atmosphere, and the temperature. In general, the oxidation proceeds at a faster rate as the temperature is increased. With metals such as iron (steel) and copper, the oxide layer continues to get thicker with prolonged exposure, although the rate of oxidation slows down with time (fig. 1.7). On the other hand, the oxide layers on aluminium and on metals containing chromium reach an optimum thickness. At this point the oxide layer becomes impervious to oxygen, which cannot penetrate to the underlying metal. Further oxidation is thus prevented, and the oxide layer remains at a constant thickness. Steels containing 20% to 25% chromium are used

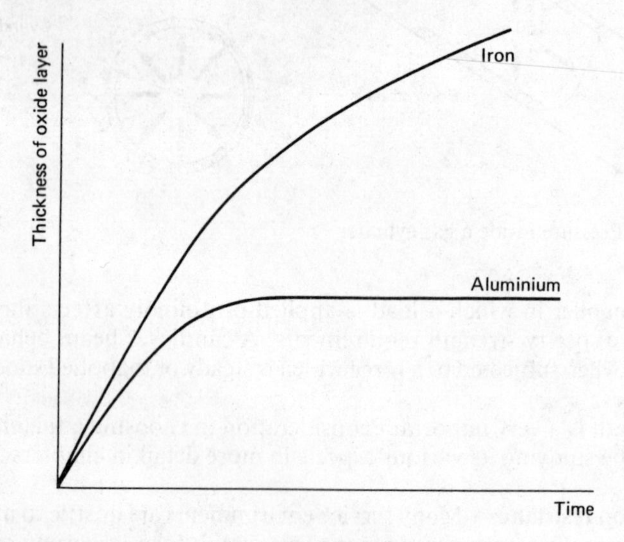

Fig. 1.7 Growth of oxide layers on iron and aluminium

for furnace linings because they have good oxidation resistance at high temperatures.

Probably the most common form of corrosion occurs when a metal is in contact with water which has small amounts of chemical dissolved in it. The mechanism is similar to the reaction which takes place in a simple galvanic cell. In this, a strip of iron and one of copper are suspended in a dilute solution of sodium chloride (common salt) and are coupled by a wire (fig. 1.8). The iron becomes an anode (positive) and the copper a cathode (negative). Electrons flow from the iron along the wire to the copper and back through the solution to the iron. This current flow causes iron to be stripped from the anode and deposited at the bottom of the bath as iron hydroxide. This is a chemical compound – more commonly known as 'rust' – in which oxygen and hydrogen are combined with iron. The oxygen used to form the rust is taken from the solution. It is replaced by fresh oxygen absorbed from air above the surface of the liquid. As more rust is formed, the thickness of the iron anode is reduced until holes appear and the metal disintegrates – a sequence of events familiar to car owners.

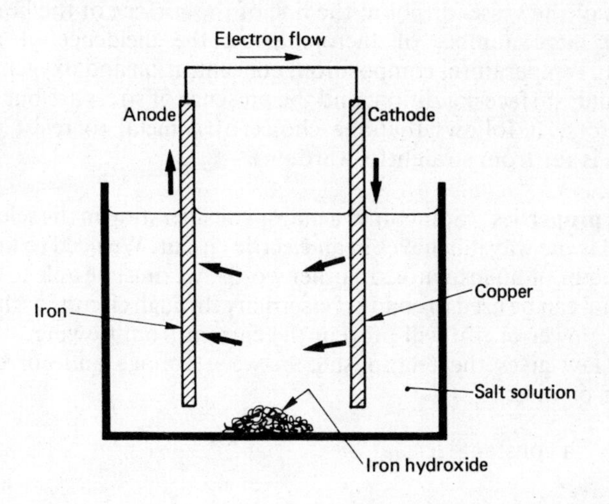

Fig. 1.8 Simple galvanic cell

Galvanic corrosion occurs in practice when two dissimilar metals are in contact and form a cell. This is why steel nuts corrode if they are used with washers made from copper or brass (a metal containing copper and zinc). Small regions of different composition on the surface of a metal can also form anodes (+) and cathodes (−), thus setting up small galvanic cells which cause corrosion. Sometimes the oxide scale on the surface of a metal can act as a cathode (fig. 1.9). In this case, localised corrosion occurs at cracks in the oxide layer. In practical situations such as these, the

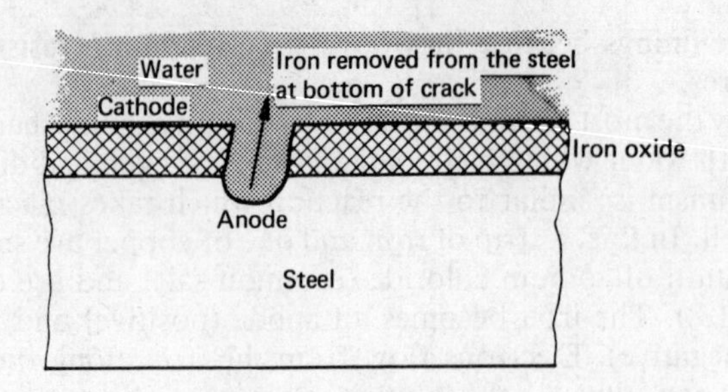

Fig. 1.9 Corrosion at a crack in an oxide layer on steel

electrolyte comes from a variety of sources. When the underside of a car rusts, the electrolyte is very often formed by moist mud which contains chemicals picked up from the road. Sea water provides the electrolyte in the corrosion of ships and the legs of oil-production platforms. In food-processing plant, dilute acids formed in the cooking operation encourage corrosion of the vessel or pot at the line of the surface of the liquid.

A very large number of factors affect the incidence of galvanic corrosion. Temperature, composition, concentration and oxygen content of the liquid; surface condition; and the presence of stress are but a few of these factors. It follows that the choice of a metal to resist galvanic corrosion is far from straightforward.

Electrical properties Sometimes a major consideration in the selection of a material is the way it behaves in an electric circuit. We need to know if it is a conductor or an insulator. In other words, we must be able to assess if the material can be used to conduct electricity through circuits with little or no loss of power or if it will prevent the current from flowing.

Ohm's law gives the relationship between voltage and current in a conductor or wire:

$$\frac{V}{I} = \text{a constant } (R)$$

where V = voltage across the conductor (volts)

I = current flowing in the conductor (amperes)

and R = resistance of the conductor (ohms)

The power dissipated in the conductor is I^2R. Hence, for good conduction we want a material which has a low resistance.

The resistance of a wire depends on its dimensions. It is more useful, therefore, to refer to the resistivity of a material, from which the resistance can be calculated:

$$\text{resistance} = \text{resistivity} \times \frac{\text{length}}{\text{cross-sectional area}}$$

Resistivity is quoted in ohm metres (Ω m), and values for some typical materials are given in Table 1.1.

Table 1.1 Resistivities of common materials

Material	Resistivity (Ω m at 20°C)
Silver	1.6×10^{-8}
Copper	1.67×10^{-8}
Aluminium	2.66×10^{-8}
Carbon steel	1.7×10^{-7}
Lead	2.1×10^{-7}
Stainless steel	7.0×10^{-7}
Graphite	1.4×10^{-5}
Alumina	1.0×10^{11}
Bakelite	1.0×10^{11}
Polythene	1.0×10^{14}
Glass	1.0×10^{16}
Polypropylene	1.0×10^{17}

In general, metals are conductors but most of them have resistivities which are too high for cables and wires in electrical circuits. Copper is widely used for this purpose, because it has a very low resistivity, but aluminium is often preferred for overhead cables as it is much lighter than copper and still has a low resistivity.

By contrast, non-metallic materials (except graphite) mostly have very high resistivities and are used as insulators. Familiar examples are the plastics coverings on cables in domestic wiring circuits and the ceramic insulators on automobile sparking plugs.

Consideration must also be given to the magnetic properties of a material. There are three common metals which are ferromagnetic, i.e. are attracted strongly by a magnet. They are iron, cobalt, and nickel. In certain circumstances the magnetic properties of the metals can be very useful. There are other applications where the use of a ferromagnetic material could lead to problems – for example, by interference with electrical equipment. In these situations it may be necessary to choose a non-magnetic material even if this may mean a compromise on other properties.

Aesthetic considerations Although in engineering we tend to look principally at the properties discussed above, it must not be forgotten that materials are also frequently chosen for their appearance. Copper and stainless steel take a high polish, the surfaces of aluminium sheets can be coated with a thick oxide layer (anodised) which can be coloured, and plastics can be printed to give decorative effects. In operations like food processing, a shiny surface to a cooking vat not only gives a good impression to the observer but is also an essential feature in the prevention of bacteria entrapment.

Manufacturing requirements

At the design stage, it is all too easy to concentrate on selecting the best combination of properties and ignore the fact that the material must be shaped, formed, or joined. With so many manufacturing methods available, it is not easy to list all the factors which must be taken into account, but the following questions indicate some of the more important points:

- is the material easy to machine?
- what type of finish is produced by machining?
- is special plant needed for casting or moulding?
- can the material be formed cold or does it need to be heated?
- how much springback is there after forming?
- are there any forming processes which cannot be used?
- is it easy to make sub-assemblies which can be readily joined together?
- can it be welded or brazed successfully?
- are heat treatments needed to achieve or restore properties?

Cost of the raw material

Clearly the cost of the raw material for the component or fabrication is of great importance. Depending on the type of product, the cost of the materials may be as much as 40% of the total manufacturing cost. The higher this percentage becomes, the greater is the attraction of finding low-cost materials. Against this, it may well be that an apparently more expensive material could be easier to manufacture, thus reducing the overall cost of the product.

Customer preference

Finally, we must not forget that the customer may impose specific conditions which have a major influence on the choice of the most suitable material. They can very often rule out the use of a material which would otherwise appear to be the logical choice. This is a very important consideration when manufacturing goods for the domestic consumer market. For example, car exhaust systems are usually made in carbon steel and then painted. In service, they rust and need replacing periodically. On the face of it, the logical course of action would be to make the exhaust systems from an alloy steel. These would have a longer life and should be more attractive to the car owner, but they would cost more. Present indications are that customers prefer the cheaper product, perhaps because they do not expect to keep their cars long enough to have to buy another silencer.

With industrial plant, the purchaser may prefer one material rather than another because facilities for repair work may be limited. On site, it is much easier to repair a crack in a carbon-steel unit, which can be readily welded by the oxy–acetylene or manual metal arc processes, than one in an aluminium alloy which would call for more specialised welding procedures. Sometimes, metal components may be preferred to cheaper plastics units because the latter cannot be repaired and replacement items must therefore be kept in stock.

2 Atomic structure

Our appreciation of materials used in everyday life is conditioned by our senses. We identify objects by their appearance, feel, smell, and taste. In doing this, we place much reliance on previous experience. Favourite foods are recognised by their taste. Smells are often associated with dangerous substances and give a warning sign. Metals look and feel solid. The piece of steel bar in the lathe chuck does not collapse when the jaws are tightened, so we deduce that it is hard. When layers are machined from the surface of the bar, the appearance of the underlying metal does not change. It would be reasonable to assume that the bar is uniform in all respects throughout its thickness.

Unfortunately it is relatively easy to draw the wrong conclusions in these assessments based solely on the evidence of sight and touch. Our eyes can give only limited information about a material. For example, all carbon steels look similar but it would be wrong to assume that they all have identical properties, as this would be far from the truth. The steel used for car bodies can be formed or pressed easily into shape, but has relatively low strength. A visually identical sheet of high-strength steel may crack when it is bent through only a few degrees. This contrast occurs because the internal structure of the two steel sheets is different. The unaided senses cannot detect variations in structure, because there is no change in surface appearance. The differences are shown up by using mechanical deformation (i.e. bending) to assess a structure-dependent property (i.e. ductility). For a more definitive comparison we could use microscopic and X-ray examination to show that there are definite relationships between structure and properties which must be understood if the best use is to be made of materials in engineering.

2.1 Atoms and the structure of matter

What does the term *structure* mean? According to the Concise Oxford Dictionary, it is the 'manner in which a building or organism or *other complete whole* is constructed. . . .'

All materials are constructed or made-up of atoms. An atom is the smallest particle of an element which can take part in a chemical reaction. ('Element', 'chemical reaction', and other basic chemical terms used in the text are explained in Table 2.1). If we took a piece of an element – say carbon – and divided it time and time again, each new portion would retain the properties of carbon. If it were technically possible to keep on subdividing, we would ultimately reach the smallest particle which could be recognised as carbon. This would be an atom. Any further division (i.e.

Table 2.1 Some chemical terms used in the text

Term	Explanation
Element	The simplest form of matter – an element cannot be separated into different substances. All the atoms in an element have the same atomic number (see section 2.3).
Compound	A substance in which two or more elements are chemically united in definite proportions. The physical properties of a compound are different from those of the elements it contains.
Molecule	Two or more atoms bonded together to form the smallest unit which retains the physical properties of the bulk substance. It may consist entirely of the same types of atom, in which case it is the molecule of an element; on the other hand, it may contain a number of different atoms to form a molecule of a compound.
Mixture	An intermingling of two or more elements or compounds. The components of a mixture are not chemically combined and retain their original identities. The physical properties are an aggregate of those of the constituents.
Solution	A special type of mixture in which the components are intermingled at an atomic level. Usually the physical characteristics of one of the components predominate, e.g. water still looks and behaves like water when sugar is dissolved in it.
Chemical reaction	Two or more elements or compounds interacting to produce chemical changes in each other.
Exothermic reaction	A chemical reaction which gives out heat.
Endothermic reaction	A chemical reaction which absorbs heat.
Oxidation	A chemical reaction which adds oxygen to a substance or removes hydrogen from it.
Reduction	A chemical reaction which adds hydrogen to a substance or removes oxygen from it.

splitting) of the atom would drastically alter its properties and the resulting particles would be unrecognisable as carbon. An alternative definition of an atom could be, therefore, that it is the smallest particle of an element which retains the physical characteristics of that element. We can visualise matter as a collection or agglomeration of atoms whose position and behaviour determine its properties and characteristics.

2.2 States of matter

Matter can exist as a gas or a liquid or a solid. In a *gas* the atoms or molecules do not occupy fixed positions but can move in a random manner. There are only very weak attractive forces trying to hold together the atoms or molecules, which can always move further apart to fill the complete volume of the vessel which contains them. Gases thus do not have either fixed volume or shape.

Atoms or molecules in a *liquid* are much closer together than in a gas and are lightly bonded to each other. They possess appreciable energy and can move past each other with a sliding motion. As with gases, a liquid does not have a fixed shape. It can take up the shape of a vessel or container, but its volume remains constant. Although the atoms or molecules are in contact with each other, they do not take up fixed positions; in other words, there is no order in the arrangement of the atoms of a liquid. Liquids are therefore said to be *amorphous* (literally, 'shapeless' or 'unorganised').

Solids are recognisable by the fact that their atoms occupy fixed positions. The strong bonds which exist between them prevent the sort of movement we have seen in gases and liquids. The atoms are located at sites which are decided by the structure of the atoms and the nature of the bonds. The combination of the properties of the atoms and the way in which they are assembled into collections determines the essential characteristics of a material. The important difference between metals and plastics is that in metals there is a degree of order in the atomic arrangement which is absent in plastics.

Metals are made up of a large number of atoms joined together in an orderly fashion known as a lattice structure. The mechanical properties of a particular metal are related to the patterns found in its lattice. By contrast, the whole range of plastics which have done so much to revolutionise modern life is based on various ways of combining carbon, hydrogen, and oxygen atoms to produce molecules which have characteristics quite different from any of these elements.

Atoms are very small: an iron atom has a diameter of 1.24×10^{-10} m. A piece of iron 1 mm^3 in volume contains many millions of atoms.

In discussing the atomic structure of a piece of iron, we are looking to see if there is a clearly defined pattern of groupings which is continued throughout the whole collection of atoms. Stated more precisely, we are looking for evidence of long-range order. Iron, in common with other metals, has a structure in which the atoms are arranged in regular lines, in layers throughout the thickness (fig. 2.1). Materials which show long-range order are termed *crystalline*. This type of structure is not confined to metals – some non-metals such as common salt (a compound of sodium and chlorine) and graphite (a form of carbon) also have a crystalline structure.

However, most non-metals are not crystalline. There may be a total absence of any recognisable repeated grouping. An example is glass, which is composed of silicate molecules distributed in random order.

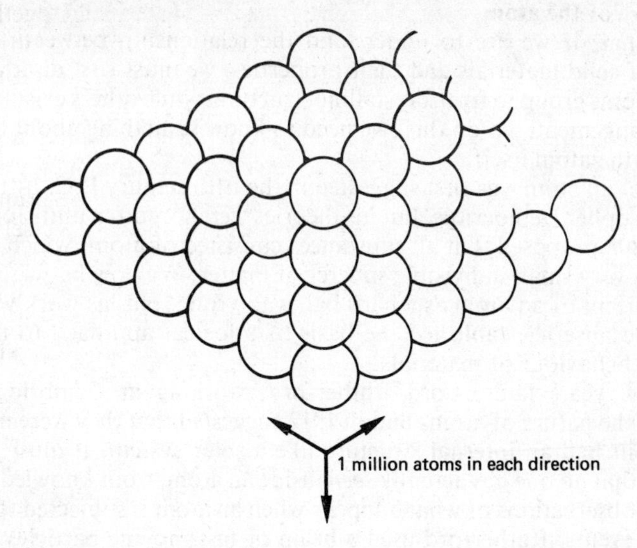

Fig. 2.1 Layers of atoms in iron

Usually there is some short-range ordering in materials such as polymers. In these, atoms of carbon and hydrogen combine to form long chains (fig. 2.2). Each chain is made up by linking together a number of basic units called *mers* in which the hydrogen atoms occupy fixed positions in relation to the carbon atoms. There is thus a short-range order within each chain.

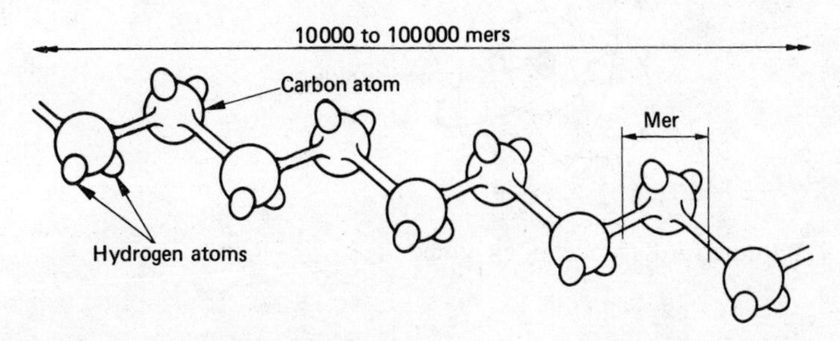

Fig. 2.2 Molecular chain of polyethylene

Materials which do not show long-range order are often termed non-crystalline. They can also be regarded as amorphous, since the atoms or molecules are collected together in a random order. In this, they show some similarity with liquids but, of course, they differ in the essential respect that in solids the atoms do not move from their allocated positions.

2.3 Structure of the atom

It is clear that, if we are to understand the relationship between the structures of solid materials and their properties, we must first discover why some atoms group to form crystalline structures while others exist in a random arrangement. To do this, we need to know something about the structure of the atom itself.

The idea of an atom was first suggested in the fifth century B.C. by the Greek philosopher Democritus, but his theories were forgotten until John Dalton in 1808 proposed that all substances consisted of atoms which he visualised as very small indivisible spheres of matter. We now know that Dalton's concept of an atom as a hard ball was wrong, but his work was important because it established the basis of a logical approach to the study of the behaviour of materials.

About 100 years later, Lord Rutherford, working at Cambridge, investigated the nature of atoms and in 1912 suggested that they were not indivisible but had an internal structure like a solar system. It must be appreciated that no one has actually *seen* inside an atom – our knowledge comes from observations of what happens when an atom is subjected to a given set of events. Rutherford used a beam of fast-moving particles to bombard a piece of metal foil and then observed the way in which they were deflected. These particles – known as α particles (*alpha* particles) – have a positive charge. They are smaller than atoms and are scattered in different directions if they collide with an atom. By analysing the scatter patterns, Rutherford was able to produce a model of the internal structure of an atom.

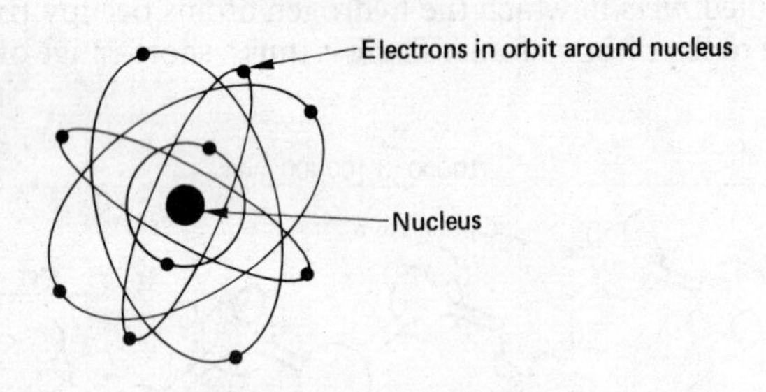

Fig. 2.3 Rutherford's model of an atom

Rutherford's model (fig. 2.3) consisted of a positively charged *nucleus* surrounded by negatively charged *electrons* which moved along specific orbits. In a sense, he visualised a miniature version of our solar system in which the planets orbit the sun. There are a number of orbits at different distances from the nucleus. Electrons moving within these orbits form shells around the nucleus. Since the atom is electrically neutral, the total negative charge on the electrons in all the shells must be equal to the positive charge on the nucleus.

Subsequent work by atomic physicists, in particular Niels Bohr, has lead to a model of the atom in which the electrons are seen to be in a cloud, called an *orbital*, rather than moving along fixed paths. There are a number of orbitals corresponding to the shells suggested by Rutherford. For our present purposes, the exact shape of the orbital is not important – we can view it as a zone of energy in which the electrons can exist. This zone is almost empty, but the electrons move so fast that it gives the appearance of being full. We can regard the diameter of the outer orbital as being the size of the atom.

Nucleus of an atom

The mass of an atom is concentrated in the nucleus: the electrons make only a small contribution to the total mass. The nucleus contains two principal types of particle: (a) positively charged *protons* and (b) *neutrons* which are electrically neutral, i.e. they have no charge. The charge on the proton is equal in magnitude to that on the electron but is of opposite sign; i.e. a proton is positive whereas an electron is negative.

The number of protons in the nucleus is a characteristic of the element and is known as the *atomic number*. Iron has an atomic number of 26, i.e. there are 26 protons in the nucleus and hence 26 electrons in the shells. Carbon, on the other hand, has only 6 protons in the nucleus, while lead has an atomic number of 82.

What is the role of the neutrons? Although they do not alter the electrical charge on the nucleus, they do contribute to the mass of the atom. The *atomic mass number* (A) is the sum of the numbers of protons (Z) and neutrons (N) in the nucleus:

$$A = Z + N$$

Iron has an atomic mass number of 56. Its atomic number, Z, is 26; hence there are $56 - 26 = 30$ neutrons in the nucleus. Similarly for

carbon	$A = 12$	$Z = 6$	$N = 6$
copper	$A = 64$	$Z = 29$	$N = 35$
lead	$A = 207$	$Z = 82$	$N = 125$

Although the atomic number (i.e. the number of protons) remains fixed for a given element, the number of neutrons can vary, giving different atomic masses. These variants are known as *isotopes*. For example, as normally bought from the metal supplier, magnesium has an atomic number of 12 and its atomic mass is 24; hence there are 12 neutrons in the nucleus. There are also isotopes which have 11 and 13 neutrons, giving atomic masses of 23 and 25 respectively. These are not normally of importance in general engineering practice – when we make a magnesium casting we are using the metal which has an atomic mass of 24. However, the isotopes of some metals are unstable and emit γ rays (*gamma* rays) as they change to the normal form. These rays are capable of penetrating metals and can be used in the same way as X-rays to examine castings, forgings, and welded joints. Iridium-192, cobalt-60, and caesium-137 are examples of radiation sources commonly used in commercial practice – the number indicates the atomic mass and thus identifies the isotope being used.

The electrons

The mass of an electron is only $\frac{1}{1836}$ of that of a proton. In spite of this, the electron possesses the same size of charge as a proton, but of opposite sign. It follows that in an atom which is electrically neutral there are an equal number of protons and electrons. We saw above that the electrons in an atom are arranged in orbitals or shells surrounding the nucleus. In the atoms of the elements which we know about, there can be up to seven shells. The actual number and the distribution of electrons between shells is related to the atomic number (Z) of the element, although the first orbital or innermost shell always contains only two electrons (except for hydrogen, which has only one electron altogether). The higher the number, the more shells will be present. Hydrogen ($Z = 1$) has only one shell, whereas iron ($Z = 26$) has four and uranium ($Z = 92$) has all seven.

2.4 Covalent bonding

A significant feature of the structure of the atom is the number of electrons in the outermost shell. These are called the *valence electrons* and they are important in determining the ability of an atom to bond with other atoms. The number of electrons present in the outer shell or orbital is a characteristic of the element and, in general, is from one to eight. For example, sodium has one electron in the outer shell, magnesium has two, aluminium three, carbon four, and oxygen six. The outer orbital is completely full when it contains eight electrons (e.g. in argon or neon). If there are less than eight electrons, we can think of the outer orbital as being incomplete and therefore containing vacant sites. This is an important concept since it helps us to understand chemical bonding.

In most non-metallic compounds, atoms are joined together by *covalent bonding*, which is a sharing of valence electrons. We will be studying this type of bonding in greater detail when we discuss the structure of plastics.

In a sense, covalent bonding can be viewed as an intersection of the outer orbitals of two atoms. Consider two atoms, one of which has two vacancies in the outer shell while the other has six, i.e. it has only two electrons in the orbital. Wherever possible, atoms will try to complete the outer shell and so achieve the most stable condition (fig. 2.4). If these two atoms are brought into contact, they can share the two valence electrons so that each has a potential of eight electrons in its outer orbital. The shared electrons revolve around both atoms.

Two or more atoms bonded together form a molecule. If the atoms are of the same kind, the molecules will have the same physical properties as the atom. For example, a molecule of hydrogen contains two atoms of this element and has the physical properties we associate with the element hydrogen. On the other hand, one atom of carbon (solid), four atoms of hydrogen (gas), and one of oxygen (gas) can combine to form a molecule

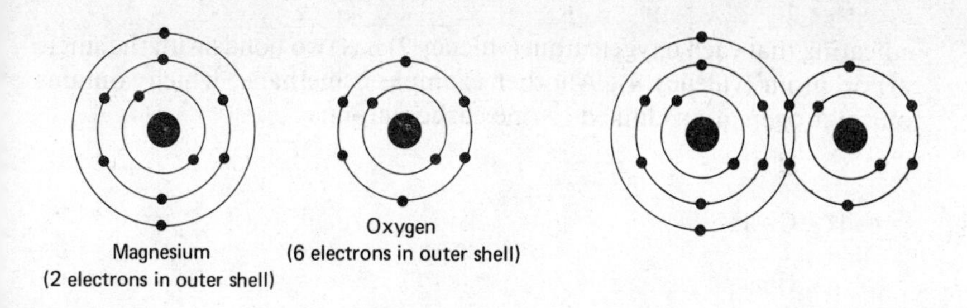

Fig. 2.4 Covalent bonding between magnesium and oxygen

of methanol (methyl alcohol) which is a liquid at normal temperature and pressure and has completely different characteristics from any of the three constituent elements.

Where the outer shell of an atom is complete, containing eight electrons, there are no electrons available for sharing and the atom is stable. It will not form bonds with other atoms. The elements which fall into this category are collectively known as the inert gases, as they will not enter into chemical reactions with other elements. Helium and neon are two well known inert gases.

Covalent bonding in organic compounds

The number of electrons or vacancies in the outer or valence shells which are available for sharing with other atoms determines the *valency* of a particular element. The valency indicates the probability that an atom will form a strong bond. Hydrogen has only one shell, which contains one electron. It has a valency of 1 and is unlikely to join with more than one other atom at a time. Carbon, however, has a valency of 4 – it could join with four hydrogen atoms or with two oxygen atoms (valency 2). Alternatively, two carbon atoms could join together while each was also bonded to two hydrogen atoms.

In later chapters of this book we will need to show the structure of a molecule to explain how it affects properties. Molecules are, of course, three-dimensional and are difficult to depict on a two-dimensional page. We therefore adopt a convention in which an atom is indicated by the chemical symbol of the element (see Table 2.2) and the number of valency bonds are shown by straight lines between the symbols. Using this convention, the hydrogen molecule is shown thus:

H – H

i.e. a single bond between two hydrogen atoms. Similarly, carbon dioxide is

O = C = O

indicating that each oxygen atom (valency 2) has two bonds with the single carbon atom (valency 4). Another example is methane, which contains four hydrogen atoms linked to one carbon atom:

```
    H
    |
H – C – H
    |
    H
```

For a molecule to be stable, all the valency bonds of the atoms involved must be used. Where bonds are available, the molecule will try to seek further atoms with which to combine.

It is important not to confuse valency and bond strength. The valency tells us the number of bonds which can be formed with other atoms. It does not mean that the link between oxygen and carbon which has two bonds is necessarily stronger than that between the carbon atom and each of the four hydrogen atoms (see also page 58).

Table 2.2 Some chemical elements and their symbols

Element	Symbol	Element	Symbol
Aluminium	Al	Manganese	Mn
Antimony	Sb	Molybdenum	Mo
Arsenic	As	Nickel	Ni
Cadmium	Cd	Nitrogen	N
Carbon	C	Oxygen	O
Cerium	Ce	Phosphorus	P
Chlorine	Cl	Silicon	Si
Chromium	Cr	Sodium	Na
Copper	Cu	Sulphur	S
Fluorine	F	Tin	Sn
Hydrogen	H	Titanium	Ti
Iron	Fe	Zinc	Zn
Lead	Pb	Zirconium	
Magnesium	Mg		

On page 17 we referred to *mers* as the basic unit for a whole range of plastics. The structure of one type of mer is:

```
  H
  |
– C –
  |
  H
```

It has two carbon bonds unused. If we add two more hydrogen atoms we produce methane, but by joining it to other mers we create a chain:

```
  H  H  H  H
  |  |  |  |
···-C-C-C-C-···
  |  |  |  |
  H  H  H  H
```

These chains can be very long, containing between 10^3 and 10^5 mers. If the first and last mers are rendered stable by adding a hydrogen atom to the unused bond, a plastics material known as polyethylene is formed:

```
  H  H                                H  H
  |  |        ten thousand mers       |  |
H-C-C-································C-C-H
  |  |                                |  |
  H  H                                H-H
```

We would not want to draw this diagram every time we wish to refer to a molecule of polyethylene. We can use a shorthand form which simply tells us how many atoms are present. Polyethylene can be written as

$$C_n H_{2n+2}$$

which states that in one molecule there are *n* carbon atoms and twice as many hydrogen atoms plus the two that seal off the ends of the chain. For polyethylene, *n* can be any number between 10^4 and 10^5. This is one example of the way in which we can construct a long-chain molecule. Other examples are given in chapter 5.

Compounds of carbon with other elements - principally hydrogen, oxygen, and nitrogen - are called organic compounds and they form the basis of the whole technology of plastics. Other well known organic materials are wood, petrol, and fuel gases. In all, about 700 000 organic compounds are known at present, and many more are being developed.

As well as the organic compounds, there are many other useful compounds formed by bonding elements together. These are termed *inorganic materials* and some are very important in engineering. Glass is a compound of silicon, oxygen, and other elements such as sodium, potassium, boron, and calcium. Concrete contains oxides of calcium and aluminium, while an essential constituent in the clay used for bricks is titanium oxide. In all of these, the basic unit is again the molecule in which atoms are held together by sharing valence electrons. For example:

```
O-Ti-O                Na       O
                        \     /
                         O-Si
                        /     \
                      Na       O
```

Titanium oxide (TiO_2)

Sodium silicate (Na_2SiO_3)

2.5 Ionic bonding

In the preceding section we discussed the most common form of bonding to be found in non-metallic materials. The essential feature of covalent bonding is the sharing of electrons. However, some atoms can enter into a different type of bond which relies on their ability to gain or lose electrons. If an electron is removed from the outer shell, the atom becomes positively charged (electropositive). When an electron is added to the outer shell, the atom acquires a negative charge (becomes electronegative). An atom which has lost or gained an electron, and therefore acquired a positive or negative charge, is called an *ion*. In this state the atoms are said to be *ionised*. Electropositive and electronegative ions attract each other and an *ionic bond* is established between them.

Sodium readily forms positive ions, and chlorine forms negative ions. In common salt, sodium and chlorine ions exist in a crystalline lattice structure (fig. 2.5). Each sodium ion is surrounded by six chlorine ions which are attracted to it. At the same time, the chlorine ions are trying to repel each other because they are of the same electrical sign. Similarly, neighbouring sodium ions are trying to repel each other. The atoms settle into a position in which all these ionic forces of attraction and repulsion balance and the system is in stable equilibrium.

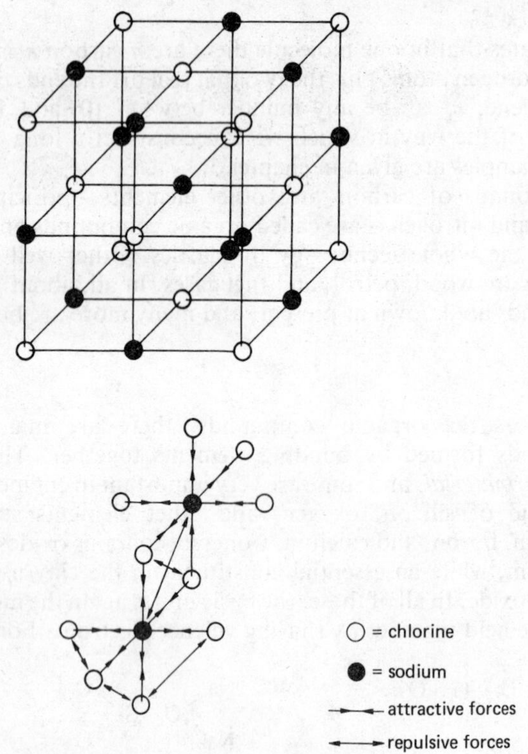

Fig. 2.5 Structure of sodium chloride (common salt)

2.6 Metallic bonding

Two thirds of all the elements are metals. Many of these will enter into chemical reactions with non-metals to form inorganic compounds in which the bonding is either covalent or ionic. These types of bond are almost exclusively found in non-metallic materials – a different, more complex, form of bond keeps together the atoms of metals themselves.

Unlike the covalent or ionic bond, the metallic bond cannot exist simply between a few atoms, say two or three or four: it is found only where there are a large number of atoms in close proximity. In a piece of metal, the valency electrons of all the atoms are shared mutually in a complex system of orbitals. In its simplest form we can visualise the structure as comprising metallic ions occupying fixed positions in a cloud of electrons which are moving along a number of limited paths (fig. 2.6). Attractive forces exist between the ions and bring them together until the outer orbitals are in contact. At this point, the positively charged metal ions start to repel each other. Their positions are fixed by the points at which attractive and repulsive forces cancel each other out. The distance between atoms when this state exists depends on the size of the atom, the number of shells, and so on and is therefore different for each metal.

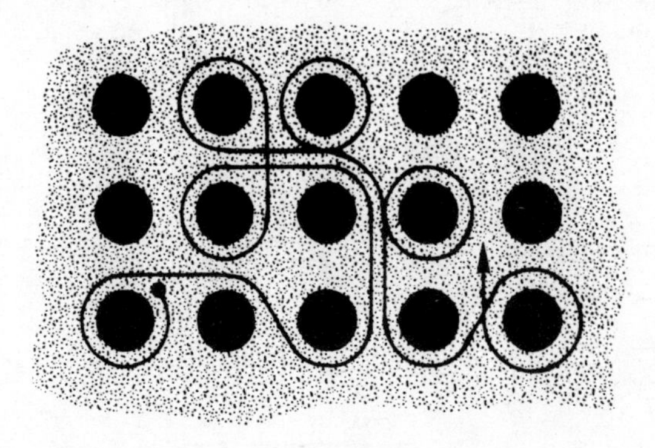

Fig. 2.6 Metal ions in an electron cloud. (The line shows a possible path of one of the valency electrons.)

The interaction between a large number of atoms leads to the formation of a three-dimensional lattice structure which is a characteristic of all metallic materials. In the majority of cases the metal atoms in a lattice occupy the corners of a cubic space lattice, but some prefer to form a hexagonal arrangement. The shape and dimensions of the lattice structure in a metal play an important role in the control of mechanical properties, and we will be exploring this aspect of materials behaviour in more detail in chapter 7.

3 Mechanical properties of materials

In chapter 1, we reviewed the many factors which a designer takes into account when selecting a material from which to manufacture a component or structure. An important consideration in the choice of a material is the way it behaves when subjected to a force. The *mechanical* properties of a material are a measure of the resistance it shows to the application of the basic types of force (fig. 3.1):

a) *tensile* force, which tries to increase the dimension of the material in the direction of the force, i.e. attempts to pull it apart;
b) *compressive* force, which tries to decrease the dimension;
c) *shear* force, which acts in such a way that one piece of the material moves relative to the other.

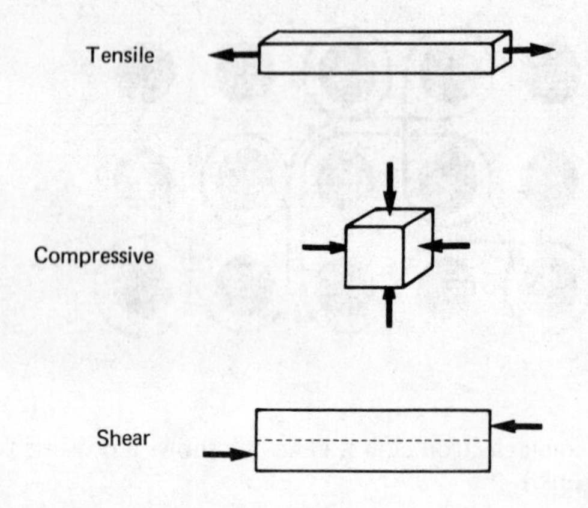

Fig. 3.1 Three basic types of force

To a large extent, in considering the nature of the mechanical properties of materials we are not concerned with *how* the force is applied, only with its type and magnitude. The situation is different when we discuss, in the next chapter, how we can measure mechanical properties. The method of application of the force then becomes significant.

3.1 Stress

When a tensile force of, say, 500 newtons is applied to a round bar of steel, the effect it has depends on the dimensions of the bar. To be more specific, it depends on the cross-sectional area. The force is uniformly distributed through the bar; hence, the smaller the diameter, the greater is the 'intensity' of the force. This intensity is called *stress* and is defined as force per unit area,

$$\text{i.e. stress} = \frac{\text{force}}{\text{area}} = \frac{F}{A}$$

If the diameter of the bar is 4 mm and it is subjected to a tensile force of 500 *N*,

$$A = \frac{\pi d^2}{4} = \frac{\pi \times (4\text{ mm})^2}{4} = 12.56\text{ mm}^2$$

$$\therefore \quad \text{tensile stress in the bar} = \frac{F}{A} = \frac{500\text{ N}}{12.56\text{ mm}^2} = 39.8\text{ N/mm}^2$$

If the same force is applied to a 2 mm diameter wire,

$$\text{stress} = \frac{F}{A} = \frac{500\text{ N}}{3.14\text{ mm}^2} = 159\text{ N/mm}^2$$

In other words, for a given applied force, stress is inversely proportional to the cross-sectional area or, in the case of a round bar or wire, the square of the diameter.

Stress is expressed in the same way for all three types of force:

$$\text{tensile stress} = \frac{\text{tensile force}}{\text{area}}$$

$$\text{compressive stress} = \frac{\text{compressive force}}{\text{area}}$$

$$\text{shear stress} = \frac{\text{shear force}}{\text{area of material in shear}}$$

3.2 Strength

If the same force of 500 N is applied to a wire which has a diameter of only 0.5 mm, the stress is 2500 N/mm^2. With a piece of low-carbon-steel wire, such as that used for fencing, the stress would never be able to reach this level as the wire would break. In everyday terms, we would say that the wire is not strong enough. How can this statement be quantified? To do this we need to introduce the concept of strength.

A piece of low-carbon-steel wire fractures when the tensile stress is raised to about 400 N/mm^2. On the other hand, a high-tensile-steel wire – for example a guitar string – can accommodate a stress of 2500 N/mm^2 before it breaks. One way of looking at the strength of a material, therefore, is the stress which is required to fracture it in tension. This is only one aspect of strength and to avoid confusion it is termed the *tensile strength* of a material.

The tensile strengths of some commonly used materials are given in Tables 3.1 and 3.2, from which it will be seen that there is a very wide range available to the design engineer.

Table 3.1 Tensile strength of some typical plastics

Type of plastics material	Tensile strength (N/mm^2)
MF resins	55–80
Nylon 6	60
Polyester resins	55
Polyethylene	10–20
Polypropylene	35
Polystyrene	50
Polyvinylchloride	10–50
UF resins	70

Table 3.2 Tensile strength of some common metals

Type of metal	Tensile strength (N/mm^2)
Aluminium	45
Aluminium–magnesium alloy	280
Brass	325
Copper	220
Low-carbon steel	410
Alloy steel	750
Zinc die-casting alloy	250

It might be asked why we should choose a material with a low strength when by using one with a very much higher strength we should be able to forget about fracture in most cases. The answer lies in the fact that tensile strength is only one aspect of mechanical properties and is rarely the sole reason for choosing a material. Consideration of the ratio of strength to weight and the behaviour at stress levels below that of fracture can be of overriding importance.

3.3 Strain

While stress is very useful in calculating the size or cross-section of the component, it is very much a theoretical concept. Indeed, we do not measure stress directly. Usually, the force and the cross-sectional area are measured and the stress is calculated from this data. Alternatively, the changes that the stress produces in the dimensions of a solid are observed. On the basis of previously assembled data, the value of stress which would be required to produce these changes is deduced. The way in which stress affects dimensions can be understood by considering what happens to our length of steel wire when the force is applied.

At the start, with no stress present, the atoms in the steel are at their stable positions. In this condition, the attractive forces trying to bring the atoms into contact are in balance with the repulsive forces created by the positive charges on the atoms. In fig. 3.2, the difference between the attractive and repulsive forces is plotted for a number of interatomic

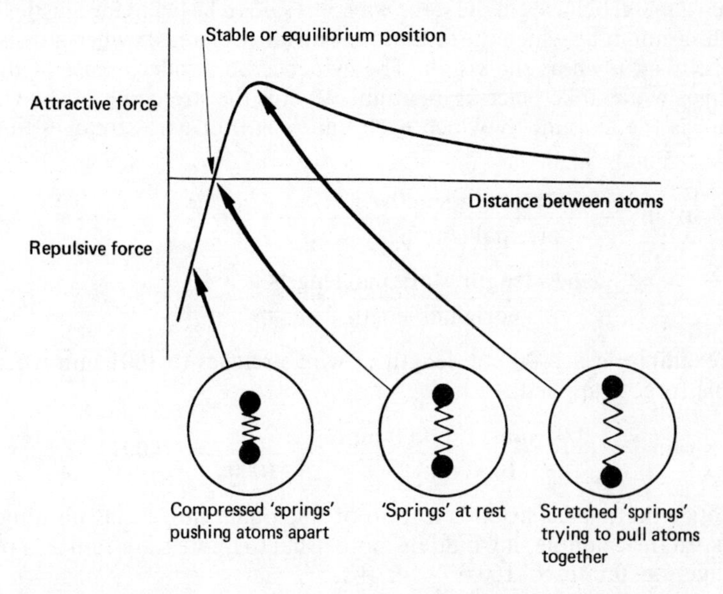

Fig. 3.2 Forces between atoms, and a simplified spring model to illustrate behaviour when compressive or tensile stresses are applied

distances. When the atoms are far apart, there are only weak attractive forces trying to pull them towards each other and the repulsive forces are negligible. As the distance decreases, the net force is still attractive, increasing to a maximum. At this point, the repulsive forces begin to be significant and the resultant force becomes less attractive until it reaches zero. With smaller interatomic distances, the repulsive forces are larger than the attractive forces and the resultant effect is that atoms repel each other. They try to move to the equilibrium position where the forces of attraction and repulsion are equal. Thus the point at which the curve for the resultant force crosses the zero line is the equilibrium interatomic distance.

If the atom is moved from its equilibrium position, forces are set-up which resist the change and try to return the atom to its original site. Such a situation exists when a force is applied to the wire. The tensile force pulls the atoms apart, acting against the attractive forces which are thus created. As long as the force is maintained, the atoms occupy a new equilibrium position. When the force is removed, the attractive force between the atoms takes over again and they are moved back to their original sites. It seems, therefore, as if the atoms in the steel are connected by a network of springs which can be stretched or compressed but which will always return to their original length when the force is removed. The property of stretching when a force is applied and returning to the original length when the force is removed is normally associated with elastic bands. When a metal behaves in the same way, it is said to be behaving elastically.

The amount by which the interatomic distance increases when a stress is applied is known as the *strain*. The evidence on a macro-scale of these changes which take place at an atomic level is the stretching of the wire. Strain is the amount by which each unit length of wire stretches under load,

$$\text{i.e.} \quad \text{strain} = \frac{\text{increase in length}}{\text{original length}}$$

$$= \frac{\text{new length} - \text{original length}}{\text{original length}}$$

For example, if a 1000 mm length of wire stretches to 1001 mm when a tensile force is applied,

$$\text{strain} = \frac{1001\text{ mm} - 1000\text{ mm}}{1000\text{ mm}} = \frac{1}{1000} = 0.001$$

Notice that, since strain is a ratio of like quantities, it has no dimensions. In the example, it would be more usual to quote the strain as a percentage, i.e. $0.001 \times 100\% = 0.1\%$.

3.4 Relationship between stress and strain

It is obvious that, as the stress increases, the wire stretches more. The relationship between stress and strain is given by Hooke's law. This states that in a spring or a material which behaves in an elastic manner, the extension or stretch is proportional to the force applied. This can be rephrased as stress is proportional to strain, or

$$\frac{\text{stress}}{\text{strain}} = \text{constant}$$

Hence, plotting stress against strain gives a straight-line relationship (fig 3.3).

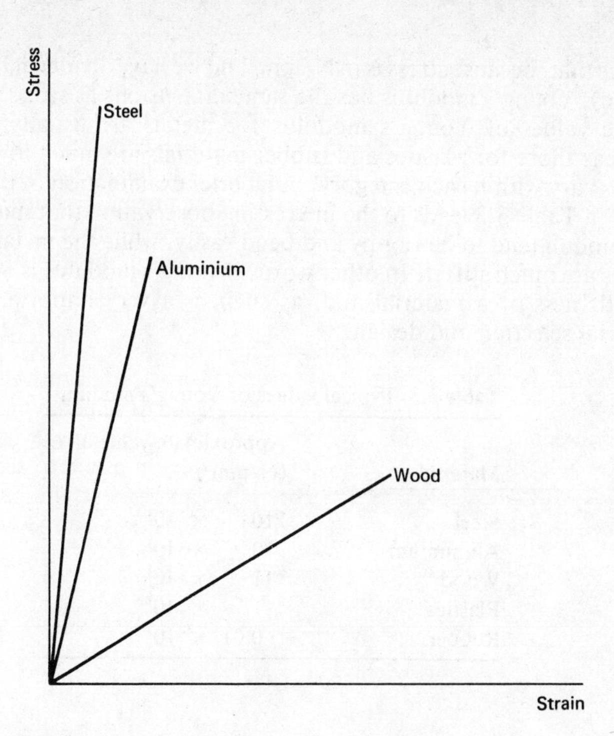

Fig. 3.3 Stress-strain relationship for typical materials

Young's modulus

The ratio of stress to strain is constant for a given material, but if we plot stress–strain curves for a number of different materials it will be seen that the slope of the line alters from one to another. This is the same as saying that the value of this ratio is a unique characteristic of the individual material. The ratio is called Young's modulus of elasticity, but it is also referred to simply as the elastic modulus.

$$\text{Young's modulus } (E) = \frac{\text{stress}}{\text{strain}}$$

Returning to our steel rods and wires, a force of 500 N applied to a bar of 2 mm diameter gives a stress of 159 N/mm². This produces a strain of 0.076% or, expressed as a ratio, 0.076×10^{-2}.

$$\text{Young's modulus for steel} = \frac{\text{stress}}{\text{strain}}$$

$$= \frac{159 \text{ N/mm}^2}{0.076 \times 10^{-2}}$$

$$= 209\,210 \text{ N/mm}^2$$

Notice that, because stress is in N/mm^2 and we have divided it by a number (strain), Young's modulus has the same dimensions as stress.

The values of Young's modulus for metals are usually very large, whereas those for plastics and rubber materials are much lower. Precise values vary within each category, but a brief examination of the examples given in Table 3.3 leads to the interesting observation that materials with low moduli tend to be floppy and bend easily, while the metals with high values are much stiffer. In other words, Young's modulus is a measure of the stiffness of a material and, as such, is a very important factor in material selection and design.

Table 3.3 Typical values of Young's modulus

Material	Approximate value of E (N/mm^2)
Steel	210×10^3
Aluminium	70×10^3
Wood	11×10^3
Plastics	1×10^3
Rubber	0.01×10^3

The relevance of stiffness can be shown by considering the deflection of a simple beam when it is made of different materials. Suppose a bar, 25 mm square cross-section, is supported at two points 1 m apart and a force of 75 N is applied midway between the supports. The deflection at the mid-point is given by

$$\text{deflection} = \frac{W\ l^3}{4E\ bd^3}$$

where W = force = 75 N

E = Young's modulus (N/mm^2)

l = distance between supports = 1 m

b = width = 25 mm

and d = depth = 25 mm

$$\text{i.e. deflection} = \frac{75 \times (1 \times 10^3)^3}{4 \times 25 \times 25^3} \times \frac{1}{E} \text{ mm}$$

$$= 48 \times 10^3 \times \frac{1}{E} \text{ mm}$$

The deflections calculated for the materials in Table 3.3 are given in Table 3.4.

Table 3.4 Deflections of identical bars

Material	Young's modulus (N/mm^2)	Deflection (mm)
Steel	210 $\times$ 10^3	0.23
Aluminium	70 $\times$ 10^3	0.69
Wood	11 $\times$ 10^3	4.36
Plastics	1 $\times$ 10^3	48.00
Rubber	0.01 $\times$ 10^3	4800.00

Stress–strain curves

We have now established that there is a straight-line relationship between stress and strain when a material is behaving elastically. We also know that there is a maximum stress which produces fracture. Does the material behave in an elastic manner up to the maximum stress? How far does it follow Hooke's law? A stress–strain curve covering the complete range from zero strain to fracture provides an answer to these questions (fig. 3.4).

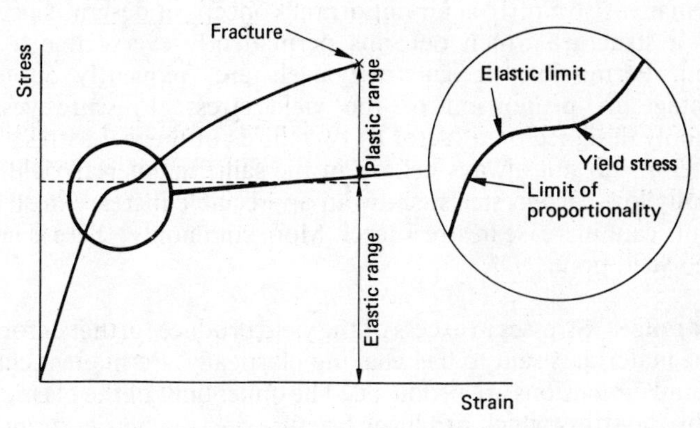

Fig. 3.4 Typical stress–strain curve

A number of significant features can be identified on a typical stress–strain curve for a low-carbon steel, as follows.

Elastic range The first part of the curve conforms to the pattern of behaviour we associate with Hooke's law.

Limit of proportionality While behaving elastically, there is often a departure from proportionality at high stress levels. This is shown as a change from a straight line to a curve. At this point Hooke's law ceases to be valid – stress is no longer proportional to strain. Even so, the material is still elastic in that it returns to its original length when the stress is removed.

Elastic limit At stress levels slightly above the limit of proportionality, the material ceases to behave elastically and some permanent deformation or elongation is produced. The elastic limit is the highest stress level to which the material can be subjected and still retain its original shape on removal of the stress.

Yield point Once the elastic limit has been exceeded, the metal yields and starts to deform plastically, i.e. it suffers dimensional changes which do not disappear when the stress is removed. At an atomic level, we are no longer in the elastic regime where the atoms move only small distances from their equilibrium positions. When the material is plastically deformed, some of the atoms move to new sites and do not return to their original position when the stress is removed. For simplicity, plastic deformation can be visualised as a sliding of one set of atomic planes over another, like a pack of cards. In reality the mechanism of deformation is more complex, and we will study it in some detail in chapter 9.

The *yield stress* is that stress required to produce a measurable amount of permanent strain. It is an important concept in design, since no one wants a structure which deforms permanently every time a force is applied. Permissible design-stress levels are frequently quoted as a percentage or proportion of the yield stress. Pressure vessels are commonly designed on a factor of two thirds of the yield stress.

Materials do not always behave in the same manner at yield. A few, especially low-carbon steels, show an appreciable increase in strain with no significant increase in stress level. More commonly, there is no clearly defined yield point.

Plastic range Stresses in excess of the yield produce further deformation, and the material is said to be behaving plastically. Permanent changes in shape and dimensions are produced. The upper limit of the plastic range is fixed by the stress which produces fracture, i.e. the tensile strength.

There is no readily defined relationship between stress and strain such as we observed in the elastic range – the amounts of plastic strain produced by increases in stress depend critically on the structure of the material. With metals in particular, the shape of the stress–strain curve in the plastic range is drastically affected by mechanical working, e.g. rolling, stretching, and forming by pressing or bending.

Ductility The amount of plastic strain before fracture is a measure of the ductility of the material. Ductility is difficult to define in precise terms. A ductile material is characterised by the ability to accommodate large amounts of plastic strain or permanent deformation before the tensile strength is reached. This property is important in manufacture. Ideally, metal sheets which are to be formed by pressing should have a substantial plastic range, probably associated with a reasonably low yield stress. This enables complex pressings to be made in one operation.

3.5 Creep strength

When we discussed the elastic behaviour of materials, it was stated that there is no permanent deformation after the stress is removed. One way of distinguishing elastic strain from plastic strain is by the ability or inability of the material to return to its original dimensions. There is the possibility, however, that some permanent strain can be induced even within the elastic range. This phenomenon is known as *creep* and is particularly observed in metals operating at elevated temperatures. Pipes carrying hot liquids or gases in chemical-processing plant can sag between supports, even though the stress level is below the elastic limit. Turbine blades in jet engines become longer under the effect of centrifugal forces and, if left for long periods without checking, can cause damage by fouling the surrounding casing. Lead sheets used to line chemical-processing vats can creep under their own weight and become thicker at the lower levels. Some plastics containers permanently change shape even at room temperatures.

Creep takes place over long periods of time and may be almost imperceptible to the casual observer. A plot of strain (i.e. deformation) against time shows that creep progresses in three stages (fig. 3.5). (As the time scale would otherwise be unmanageably long for some materials, we quite often plot a curve of strain against logarithm of time.) The first stage consists of a short period during which strain increases rapidly. There is then a long period where the rate is much slower and is reasonably constant. Finally, in stage 3, the creep rate increases and the material fractures.

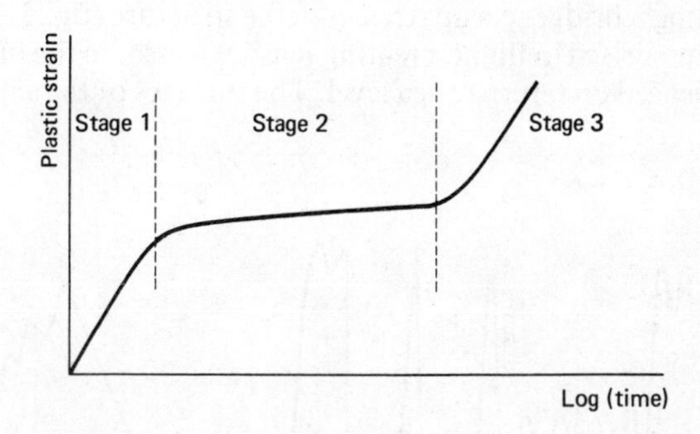

Fig. 3.5 Typical creep curve

Whether or not stage 3 exists depends on the stress level. By plotting strain–log(time) curves for a number of stresses, it can be seen that there is a critical stress level (fig. 3.6). Below this, stage 3 is absent and, although creep occurs, the material will presumably not fracture. Above the critical

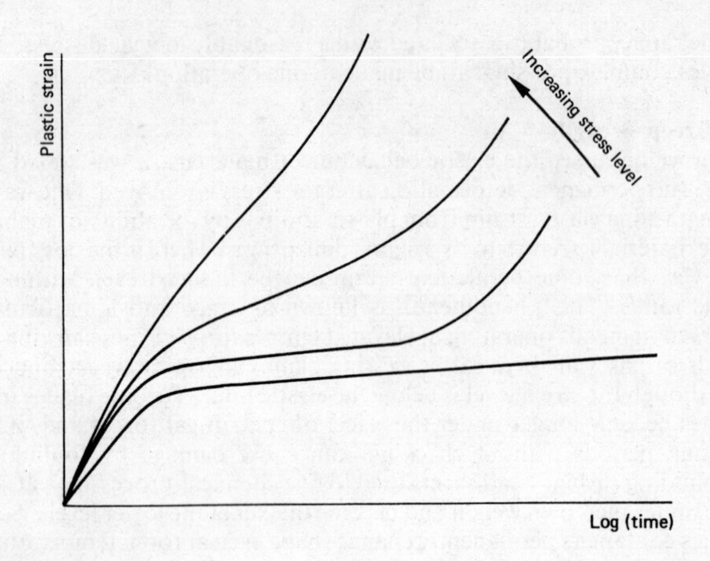

Fig. 3.6 Effect of stress level on the shape of creep curves

value, fracture occurs within an increasingly shorter time as the stress level is raised nearer to the elastic limit.

3.6 Fatigue strength

In everyday life, materials are rarely subjected to a single force which does not change, i.e. a static load. Generally the load varies in magnitude. Each vehicle crossing a bridge sets up stresses in the structure (fig. 3.7). Aircraft cabins are pressurised in flight, creating tensile stresses in the fuselage skin which are released on return to sea level. The surfaces of the axle shaft of a

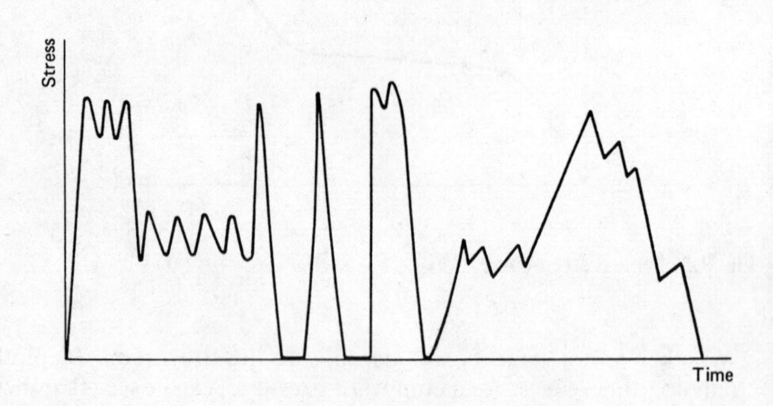

Fig. 3.7 Variations in stress experienced by a bridge as vehicles cross it

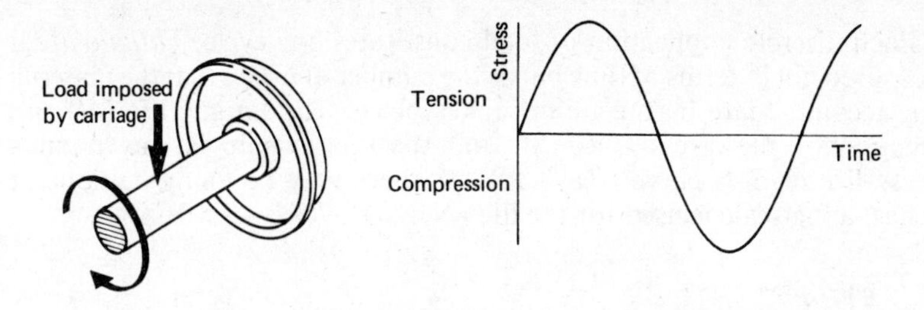

Fig. 3.8 Alternate tension and compression in a railway axle

railway carriage are alternately put in tension and compression as the wheel rotates (fig. 3.8).

In each of the examples mentioned, failures have occurred in service in spite of the fact that the stresses involved were always below the elastic limit. Large cracks have been discovered in the steelwork of motorway bridges. The disastrous failures of the first jet airliners (the Comet) were associated with cracks formed at the corners of window openings. Railway axles have to be continually checked using ultrasonic flaw detection to monitor the onset of cracking.

The important point to note in each of these cases is that the stress levels were not constant or static. They fluctuated from zero to tensile, or from zero to compressive, or they alternated between tensile and compressive. Under such conditions of dynamic loading, fatigue cracks are propagated through the material, presenting a very real risk of total failure.

Three stages can be identified in a fatigue failure: firstly the initiation of the crack, then propagation, and finally complete fracture. Fatigue cracks start at the surface, and each application or change of load extends the crack by a small amount. As the crack grows, the amount of solid material left to carry the load decreases. The stress in this region increases until simple tensile or compressive failure occurs. The surface of the fracture clearly shows the smooth region of the fatigue crack and the coarser crystalline fracture (fig. 3.9). The origin of the crack can be identified, and the smooth region often contains semicircular markings (beach marks) which correspond to the stepwise progression of the fatigue crack.

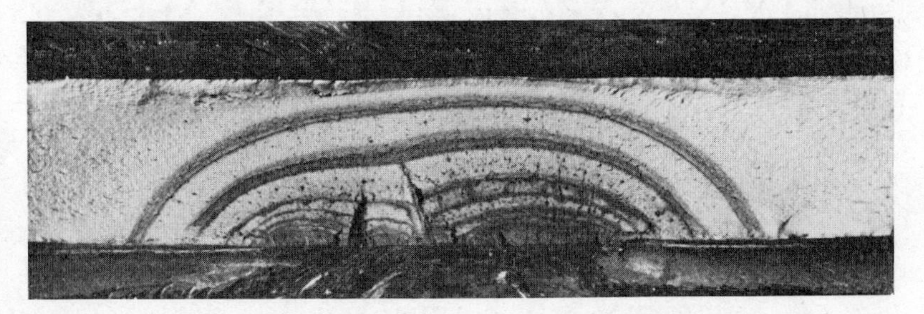

Fig. 3.9 Fracture surface of a fatigue failure

Each discrete application of load constitutes one cycle. *Fatigue life* is measured not in terms of time but as the number of cycles that the material can accommodate before failure takes place. The fatigue life (N) is a function of the stress level (S), and the relationship is shown on a stress–life or S–N curve (fig. 3.10). As there may be many millions of cycles, a log scale is used for the life (N).

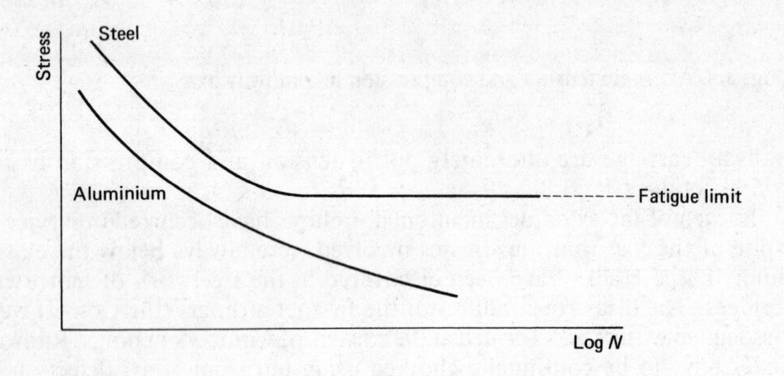

Fig. 3.10 *S–N* curves for steel and aluminium

Materials can be divided into two groups according to the shape of the S–N curve. In both groups the life increases markedly as the stress is lowered. With materials such as steel, there is a stress level below which fatigue cracks are not initiated and the life is indefinite. By contrast, this fatigue limit is not found with aluminium alloys and thermosetting plastics, for which a definite life exists even at low stress levels.

The S–N curve is also affected by the presence of notches. These concentrate the stresses at a point, so making it easier to initiate the fatigue crack (fig. 3.11). This effect can be interpreted in two ways: either the life is shorter because fewer cycles are needed to start the crack, or the

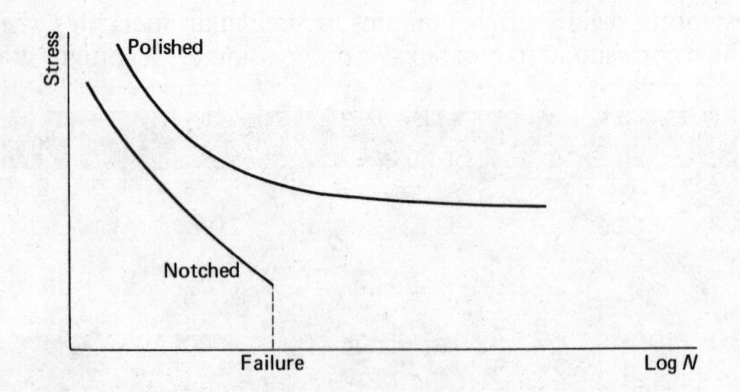

Fig. 3.11 *S–N* curves for polished and notched steel bars

maximum stress for a given life is lower. Both sharp corners and deep scratches act as stress raisers, and great care must be exercised in the design and manufacture of components subjected to dynamic loading, to minimise the chance of initiating fatigue cracks.

What, then, is the fatigue strength of a material? As we have seen, some metals have a fatigue limit, but there is no unique value for fatigue strength. The answer must be related to service conditions and requirements. Once the designer has specified the desired life, the fatigue strength can be regarded as the highest stress which will give the required number of cycles.

4 Mechanical testing of materials

4.1 Reasons for mechanical testing

There are four main reasons for mechanically testing materials:

a) *Provision of design data* Once the designer has established the loading requirements of a structure or component, the mechanical properties of a suitable material can be specified. There is a need for tabulated data which enables a designer to match available materials to the specification which has been drawn up. Often this may cover little more than the yield stress and tensile strength. On the other hand, specific service situations may call for a knowledge of the hardness, Young's modulus, and ductility of a material. The latter is particularly relevant when the method of forming is being considered.

b) *Quality control in manufacture* Materials are not necessarily uniform from batch to batch – steel plates purchased to the same nominal specification or grade may differ in yield stress by as much as 50 to 75 N/mm^2, for example. Usually a specification gives a guaranteed minimum value and allows some variation above this. The tensile strength of a casting can be altered by small changes in composition and heat treatment. The properties of a component made from a thermosetting plastics are dependent on a number of processing variables. In each of these instances, mechanical tests taken at different stages of manufacture can give a guide to the uniformity and quality of the finished product.

c) *Investigation of service failures* There is always the possibility that a failure in service has resulted from the use of materials with inadequate properties. Mechanical tests can provide a useful check in these situations.

d) *Research to develop desirable properties* The requirements of modern industry are continually presenting demands for new materials. Research workers make great use of standard mechanical tests but have also developed sophisticated techniques to provide specialised information.

4.2 Test-pieces

Ideally, the complete component should be mechanically tested to ensure that the properties of the finished product satisfy the customer's requirements. This is rarely feasible as, by its nature, mechanical testing is destructive. Our scope for product checking is limited to hardness measurements, which do not damage the component, or destructive testing of randomly selected samples in mass production.

Mechanical testing is usually carried out on test specimens cut from a piece of the material being used. Considerable care must be exercised in selecting the sample to ensure that it is truly representative of the rest of the material. Attention must also be paid to the method of preparation used for the test-piece - techniques such as guillotining, flame-cutting, and the use of high-speed slitting discs can alter the properties of the material at the cut edge and give a false result.

Standardisation of test-pieces

The results obtained during mechanical testing are influenced by the type and size of the specimen used. The elongation measured in a tensile test varies according to the gauge length (see page 42). In an impact test, different results are obtained if the shape of the notch in the specimen is changed (see page 50). If we want to compare results - especially those obtained by different test houses - the shape and dimensions of test-pieces must be standardised. Each country has its own specifications for mechanical testing, and a list of some of the relevant British Standards is given in Table 4.1.

4.3 Determination of tensile properties

In its simplest form, the tensile test involves clamping one end of a test bar while a measured force is applied to the other end. The force is increased until fracture occurs. The amount of stretching is observed and a force–elongation curve is plotted, from which values for yield stress and tensile strength are deduced. (In practice, engineers often talk of 'load–elongation' curves, using the concept of 'load' as being loosely equivalent to force. We will use force–elongation curves, but it is worth remembering that their shapes are the same as those of load–elongation curves.)

Tensile-testing machines vary considerably in size and construction (figs 4.1 and 4.2), but they all contain three essential features:

a) a means of clamping the ends of the specimen so that the force is applied axially;
b) a method of controlling the force - this is usually a hydraulic ram, but on small machines a screw thread and nut may be used;
c) a means of measuring the force.

The test-pieces used for the determination of tensile properties can be either flat or round. Flat strips can be cut directly from sheet or plate material. Round test-pieces are machined from thick plates, bar stock, and material of irregular shape such as castings or moulded components.

With both flat and round specimens, the central portion is machined with parallel sides. A gauge length is marked in this portion and is used to

Table 4.1 Selected British Standards on testing of materials
(This is not an exhaustive list of British Standards covering the testing of materials, but those mentioned amplify some of the topics introduced in this chapter. Some of them are published in several parts, and the current *BSI Yearbook* should be referred to for further details.)

Testing metals

BS 18, 'Methods for tensile testing of metals'
Definitions and symbols; form and dimensions of test-pieces; test-piece preparation and testing equipment. Procedures for determining tensile properties at room temperatures: proof stress, permanent stress, tensile stress, tensile strength, percentage elongation. Foreword gives guidance for drafting materials specifications.

BS 131, 'Methods for notched bar tests'
Izod and Charpy tests. Conditions for the test; nominal dimensions and tolerances for ferrous and non-ferrous test-pieces. Structure, dimensions of testing machine. Definitions for fracture appearance.

BS 240, 'Method for Brinell hardness test'
BS 427, 'Method for Vickers hardness test'
BS 891, 'Method for Rockwell hardness test'
Test requirements and procedure; scales and tables; minimum thickness of test-pieces; verification of testing machines.

BS 1639, 'Methods for bend testing of metals'
General principles and factors influencing test; single and reverse bend tests. Form and preparation of test-pieces.

BS 3500, 'Methods for creep and rupture testing of metals'
Definitions; test-piece dimensions; testing equipment; procedures; determination of properties and presentation of results.

Testing plastics materials

BS 5214, 'Testing machines for rubbers and plastics'
Requirements for tensile, flexural, and compression machines. Constant-rate-of-force-application machines.

BS 903, 'Methods of testing vulcanised rubber'
A series of standards dealing with the measurement of a whole variety of properties.

BS 4443, 'Methods of test for flexible cellular materials'
Measurement of tensile strength, elongation at break, cell count, compression stress–strain characteristics.

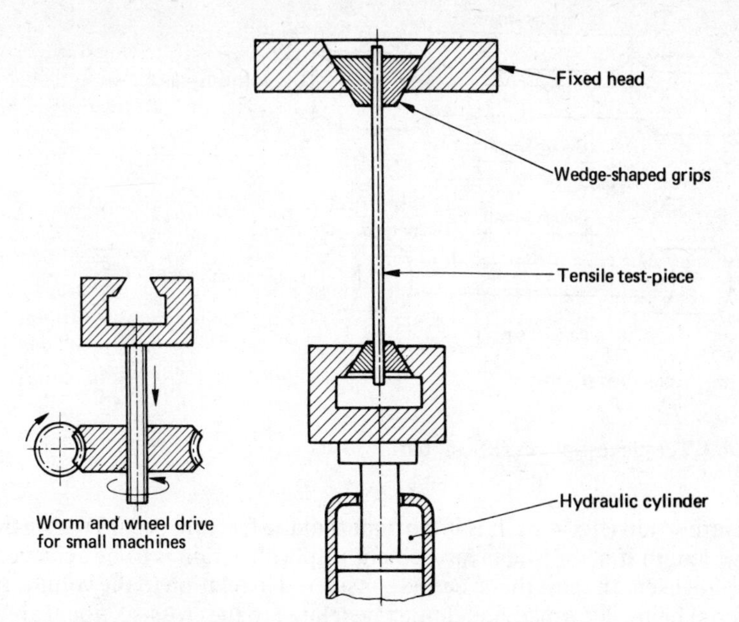

Fig. 4.1 Essential features of a tensile-testing machine

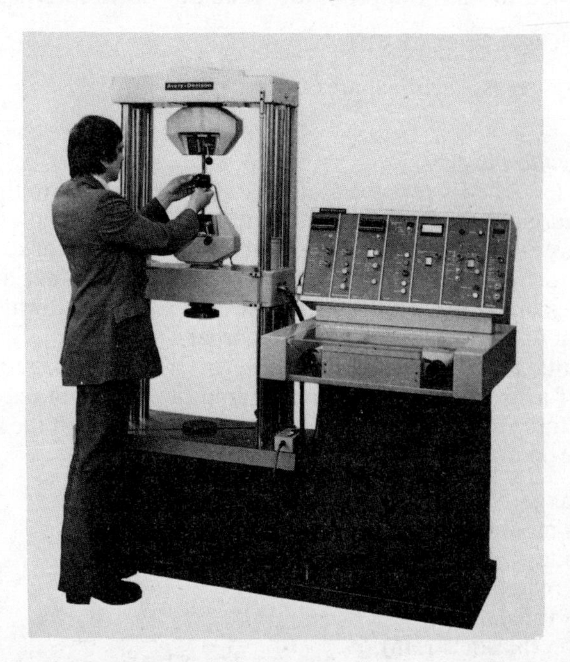

Fig. 4.2 Avery-Denison 100 kN tensile-testing machine (model 7117). In this machine the force is applied to the test specimen by two vertical leadscrews which rotate in nuts at each end of the lower cross-beam.

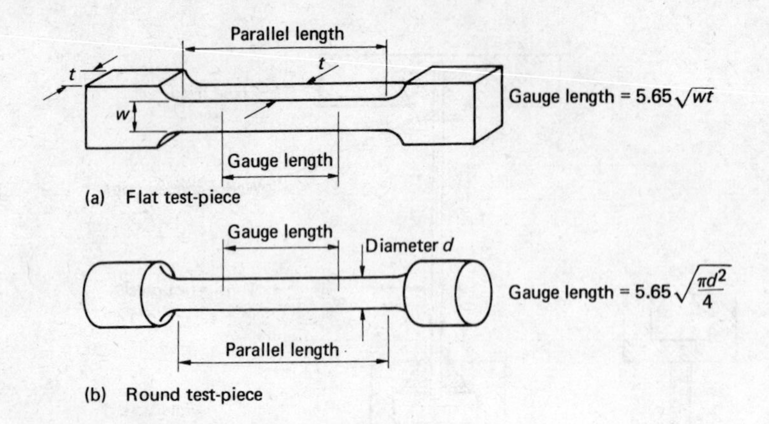

Fig. 4.3 Tensile test-pieces (BS 18: part 2:1971)

measure strain (fig. 4.3). It is important that the fracture occurs within the gauge length if a realistic measurement of plastic strain is to be achieved. The gauge length must therefore be considered in relation to the volume of material being deformed, i.e. it must be related to the cross-sectional area. This is allowed for in the appropriate standard. BS 18 specifies

$$\text{gauge length} = 5.65\sqrt{A}$$

where A = cross-sectional area of test-piece

Force–elongation curves

In a tensile test, the force is gradually increased from zero to the point at which fracture occurs. The change in the gauge length is noted for each value of force. This may be done manually using a pair of callipers, in which case the force is increased in steps to allow time to take a measurement. More commonly, an extensometer (fig. 4.4) is used which gives a reading on a dial or on an automatic recorder.

The results are plotted in the form of a force–elongation curve (fig. 4.5). If a stress–strain curve is required, it can be obtained by converting the force into force per unit area of cross-section of the specimen.

The force–elongation curve enables us to determine

a) elastic range,
b) Young's modulus,
c) yield point,
d) tensile strength,
e) breaking force, and
f) elongation (plastic strain).

In routine testing, only yield stress, tensile strength, and elongation are required.

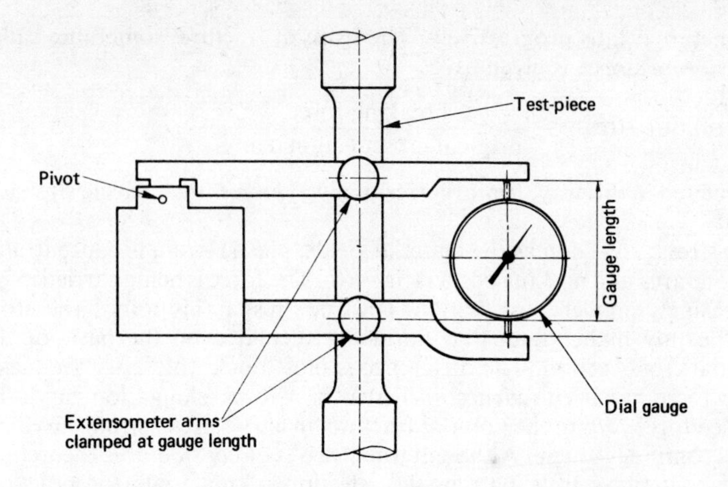

Fig. 4.4 Simple extensometer with dial gauge

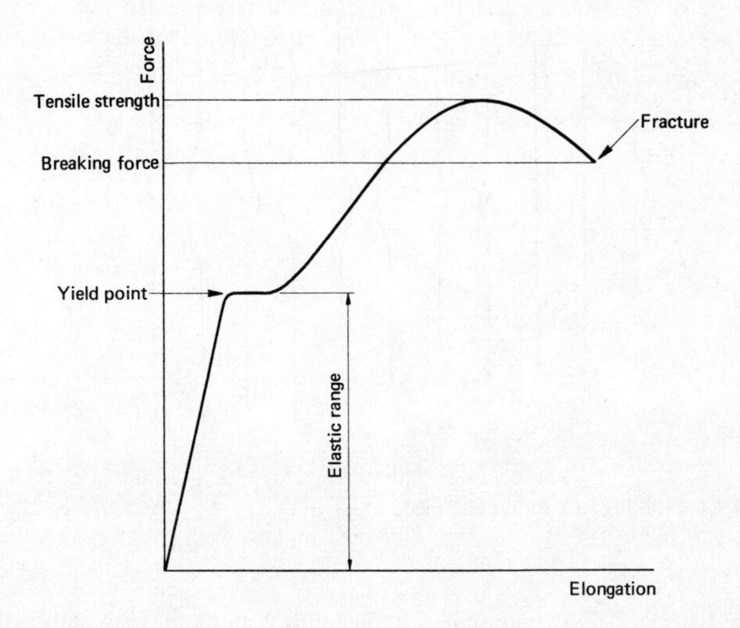

Fig. 4.5 Force–elongation curve for low-carbon steel

Breaking force The force–elongation curve does not stop when the maximum force is reached. At this point, the specimen stretches rapidly and the force required to continue deforming the specimen and ultimately to fracture it falls progressively. The stress at fracture, sometimes called the *rupture stress*, is given by

$$\text{rupture stress} = \frac{\text{breaking force}}{\text{original cross-sectional area}}$$

Calculated in this way, rupture stress is always lower than tensile strength.

True stress As soon as the material yields, plastic strain is concentrated into one area and necking occurs (fig. 4.6). The force is being carried by an increasingly smaller cross-section. The true stress at this point is therefore significantly higher than that normally calculated on the basis of the original cross-section. The difference is most noticeable after the maximum force has been reached. Whereas in a force–elongation curve the applied force *falls* to the point of fracture, in a true stress–strain curve the stress continues to *rise*. Although true stress is of considerable theoretical interest, it plays little part in the selection of materials for practical applications.

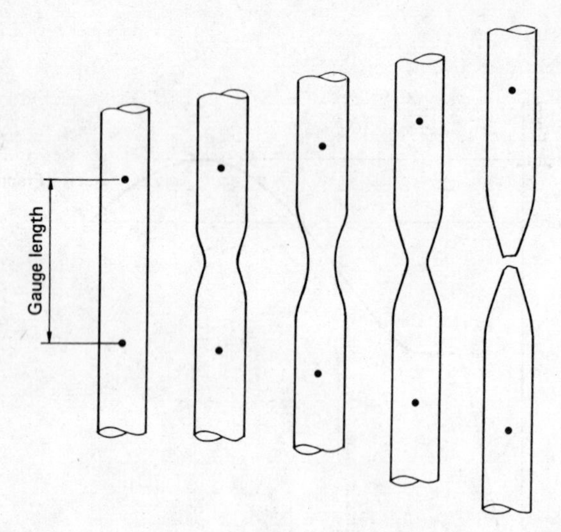

Fig. 4.6 Necking in a round test-piece

Proof stress Many materials – for example aluminium alloys and some plastics – do not show a clearly defined yield. Instead, the force–elongation curve shows a gradual transition from the straight line of the elastic range to non-linear plastic behaviour (fig. 4.7). In these circumstances it is not possible to quote a yield stress, and a convention known as *proof stress* has been adopted. Proof stress is defined as that stress which produces a predetermined amount of plastic strain, usually 0.1% or 0.2%.

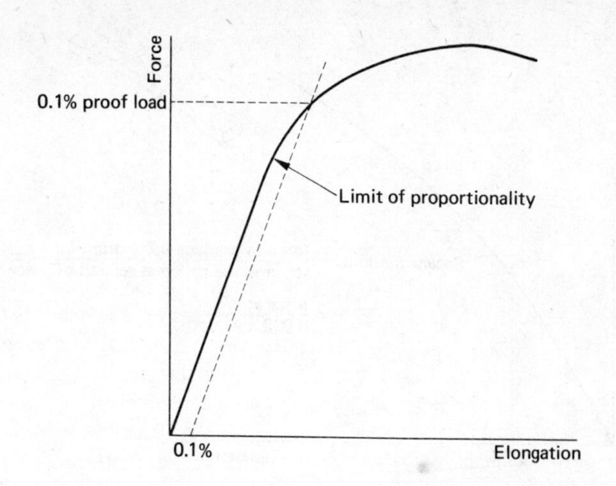

Fig. 4.7 Determination of proof stress

The magnitude of the proof stress is determined from a force–elongation curve. For the 0.1% proof stress a line is drawn through 0.1% elongation on the *x*-axis, parallel to the linear part of the curve. The line is projected until it meets the non-linear portion of the curve. The force corresponding to the intersection is the 0.1% proof load.

$$0.1\% \text{ proof stress} = \frac{0.1\% \text{ proof load}}{\text{original cross-sectional area}}$$

Secant modulus A further variation in the force–elongation curve is often observed with plastics materials (fig. 4.8). In this, there is no clearly defined straight part: the line curves slightly from the beginning. It is obviously difficult to measure the actual elastic modulus and an arbitrary value is adopted, known as the *secant modulus*.

The point on the curve at which 0.2% strain is produced *during* the test is noted. A line is drawn from this point to the origin. The slope of the line is the secant modulus for the material.

4.4 Measurement of ductility

Ductility can be defined as the amount of permanent, plastic, or elastic deformation a material can accept before rupture or fracture. It is a difficult concept to quantify, and various measurements are used to express the degree of ductility possessed by a material.

Tensile test

In the tensile test, plastic strain is expressed as permanent elongation measured over a predetermined gauge length. Percentage elongation can

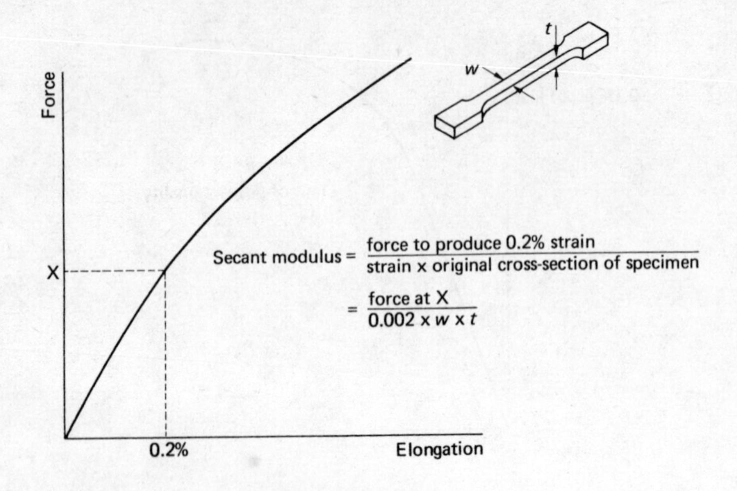

Fig. 4.8 Determination of secant modulus

be used as a measure of tensile ductility. Values of 25% to 30% would be associated with ductile metals such as pure copper and low-carbon steel, while very high-tensile-strength steels and cast iron may show as little as 5%.

With metals, the total elongation can be determined from the force–elongation curve. Where the tensile test is being used as a quality-control check, a curve is not usually plotted and the elongation is measured by placing the two broken pieces of the specimen together and noting the final gauge length.

$$\text{Elongation \%} = \frac{l_1 - l_0}{l_0} \times 100\%$$

where l_0 = original gauge length

and l_1 = gauge length after fracture

When some thermoplastic materials are being tested, there is considerable extension during the test, but much of the strain is recovered when the specimen breaks. Measuring elongation on fractured samples therefore gives a false low reading. The elongation at fracture is measured on the specimen while the force is still applied, just before fracture occurs. Alternatively, it can be read off the force–elongation curve.

Another approach to tensile ductility lies in measuring the *reduction in cross-sectional area* which occurs at the point of fracture in a tensile test (fig. 4.9). Ductile materials show appreciable necking during a tensile test, and the area of a fractured section is greatly reduced. Expressed as a percentage of the original area, the reduction in area provides a measure of

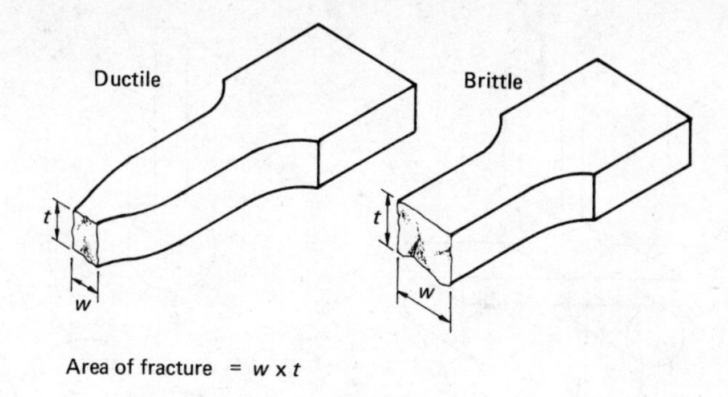

Fig. 4.9 Reduction in cross-sectional area

tensile ductility. Typical values would be 5% to 10% for brittle materials and 40% to 60% for ductile.

$$\text{Reduction of area \%} = \frac{A_0 - A_1}{A_0} \times 100\%$$

where A_0 = original cross-sectional area

and A_1 = cross-sectional area of fracture

Bend tests

There are various bend tests in common use. They all consist essentially of supporting a bar of material at its ends and subjecting it to a force at its mid-point (fig. 4.10). While it is possible to measure the onset of yielding and the amount of plastic strain on the surface which is in tension, bend tests are almost exclusively used as quality-control checks on the ductility of a material - after heat treatment, for example. In a typical test, the critical factor is the radius of the former around which the test bar is bent. A result is considered acceptable if the bar is bent through 180° without cracking. The radius of the former is specified in terms of the thickness of the test-piece. A '2*t*' test would use a former with a radius equal to twice the material thickness - it would represent a severe test of a metal's ability to be formed by bending.

Cupping tests

Ductility in sheet metal can also be interpreted as the ability to be stretched into shape. A number of cupping tests have been developed to give an assessment of this property. Essentially they involve forcing a ram into the sheet and measuring the size of cup which can be produced before the metal tears.

Probably the best-known variation is the *Erichsen test*. In this, a sample from the sheet is clamped between two rings. A round-nosed indentor is

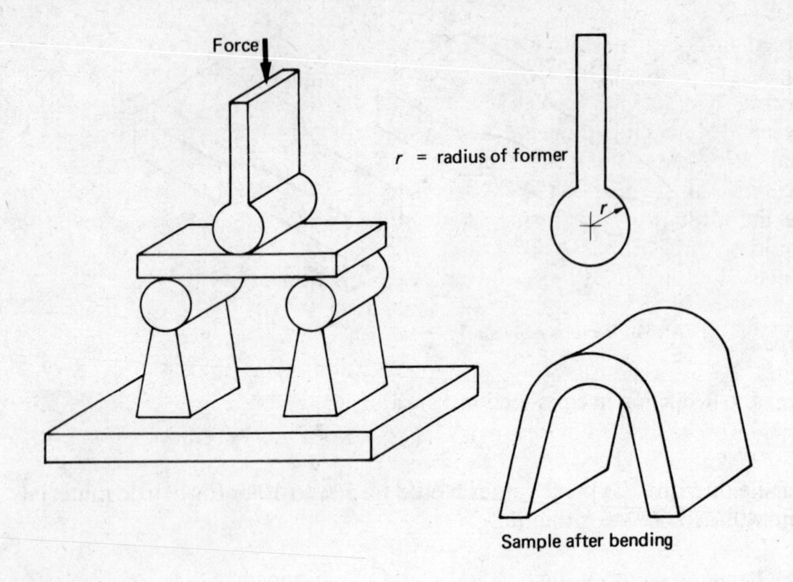

Fig. 4.10 Principles of bend tests

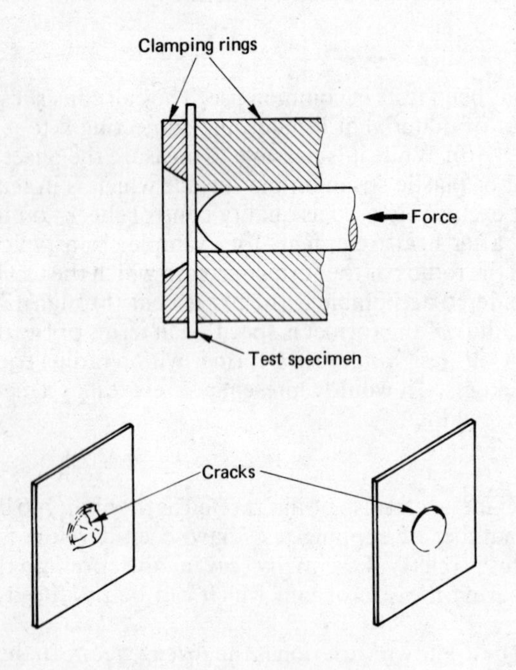

Fig. 4.11 Erichsen cupping test

forced into the sheet, which deforms to give a cup-shaped depression (fig. 4.11). The surface of the sheet is observed, and the test is stopped as soon as a crack appears. The location of the crack, its direction, the presence of necking at the crack, and the depth of the cup are used to give a qualitative assessment of the metal's ability to be stretch-formed or pressed. The surface of the sheet can also provide useful information for the manufacturer. Some metals develop an 'orange-peel' effect which would be undesirable for pressings which are to be painted, e.g. car body panels. Such a tendency would show up in an Erichsen test.

Impact tests

Brittle and ductile materials can also be distinguished by the use of an impact test. In this, a notched bar is subjected to a blow by a hammer. The test-piece is fractured by the impact, and the amount of kinetic energy lost by the hammer is reported as the *energy to fracture* or impact value.

Ductile materials absorb appreciable energy, as the area around the notch undergoes significant deformation during fracture. On the other hand, a brittle material fractures with little or no deformation and absorbs only small amounts of energy.

The machine used for the test consists of a hammer which swings about a pivot situated above an anvil holding the test-piece (fig. 4.12). The hammer is raised to a pre-set position, thus acquiring potential energy. When released, the hammer swings down to hit the specimen. It continues to swing past the anvil after fracturing the test-piece. The height it reaches is a function of the energy remaining. Hence the difference between the angles the hammer makes to the vertical before and after impact provides a measure of the energy to fracture the test-piece. With a brittle material this would be small and the hammer would swing to a high position.

There are two principal variants of the impact test: Izod and Charpy, named after the men who invented them.

In the *Izod test*, the specimen is held vertically in the path of the hammer. The force is applied at one end, and the test-piece behaves like a cantilever. The test-piece may have two or three V-shaped notches, each on a different side and 28 mm apart. These are tested in turn to allow for directionality in properties. The Izod test is used for cast iron and steel and has been used to determine the impact strength of both thermoplastic and thermosetting materials.

The *Charpy test* is now more generally preferred for both metals and plastics. It is easier to operate than the Izod test, especially at low temperatures, as the specimen is simply placed on the anvil, thus avoiding the problem of clamping. The test-piece has a single notch in the centre and is supported at each end. The notch can be in the form of a V or a U or a keyhole, but the V shape is most common.

The results of an impact test are reported in joules, although data can still be obtained in the older unit, i.e. foot pound-force (ft lbf). There is considerable scatter in the results of tests conducted on a particular material, and it is usually necessary to test three or more samples and take

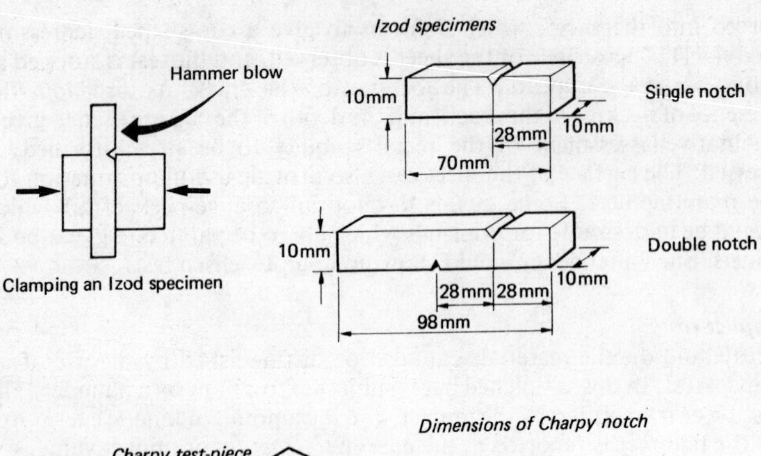

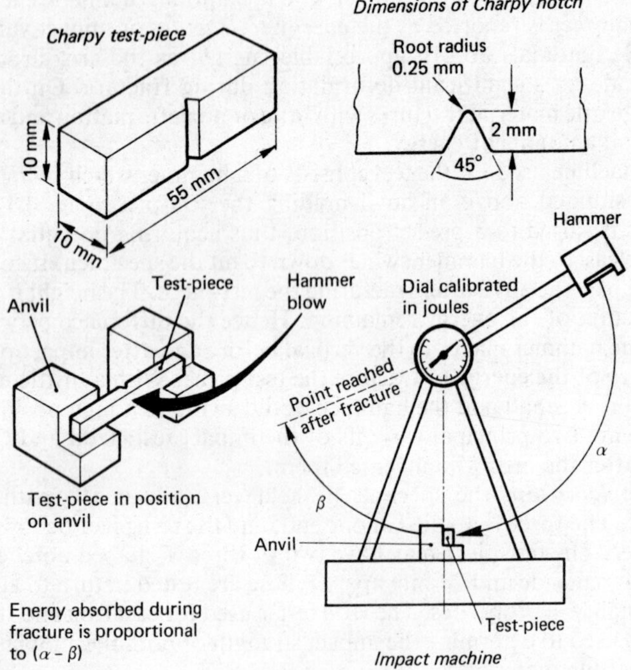

Fig. 4.12 Impact testing

an average result. The results also differ according to the type of notch used, and it is not possible to make direct comparisons between Izod and Charpy values.

Unlike tensile strength, the energy to fracture has little direct value in design – it cannot be used in calculations. Impact testing is principally used to check materials against a specification, more or less on a go/no-go basis. It is useful, for example, in monitoring heat-treatment operations such as the tempering of steels – if the tempering has not been carried out correctly, the test-piece has a low impact value.

4.5 Measurement of hardness

Hardness is not a fundamental property of a material. If asked the question 'How do you know a material is hard?', most people would answer 'Because it cannot be scratched easily' or 'It is resistant to wear' or 'It is difficult to mark the surface with a centre-punch.' To be more precise, we could say that hardness is an indication of the way in which the surface of a material deforms under specific types of localised loading.

It follows that hardness is related in some way to strength, and empirical (albeit not very accurate) formulae have been deduced to quantitatively relate hardness to the strength of a material. The main attraction in measuring hardness is that it is a useful non-destructive indicator of strength. Routine hardness measurements provide a quality-control check on the success of heat-treatment operations.

One of the earliest methods of testing was based on the concept that one material or rock could not scratch another which was harder than it. Moh's scale (Table 4.2) lists minerals in ascending order of scratch hardness. A sample of the material under test is used to make scratch marks on a series of test blocks cut from the minerals given in Table 4.2. The hardness rating of the sample lies between the Moh's number of the mineral it will just not scratch and the highest number it will mark. Moh's scale has little use in engineering apart from the testing of ceramics.

Table 4.2 Moh's scale of hardness

Moh's number	Mineral
1	Talc
2	Gypsum
3	Calcite
4	Fluorite
5	Apatite
6	Orthoclase felspar
7	Quartz
8	Topaz
9	Corundum
10	Diamond

The three major tests used to measure hardness are

a) the Brinell test,
b) the Vickers diamond test, and
c) the Rockwell test.

All three use an indentor which is forced into the surface of the test-piece (fig. 4.13). The size of the indentation is measured after the force or load has been removed, and a hardness number is calculated. With hard materials the indentation is small and the corresponding hardness number is high. Conversely, the indentation is large when soft materials are tested, so the hardness number is then low. The magnitude of the force is chosen in relation to the anticipated hardness, to give a reasonable-sized impression. Guidance is given in the appropriate British Standard (see Table 4.1).

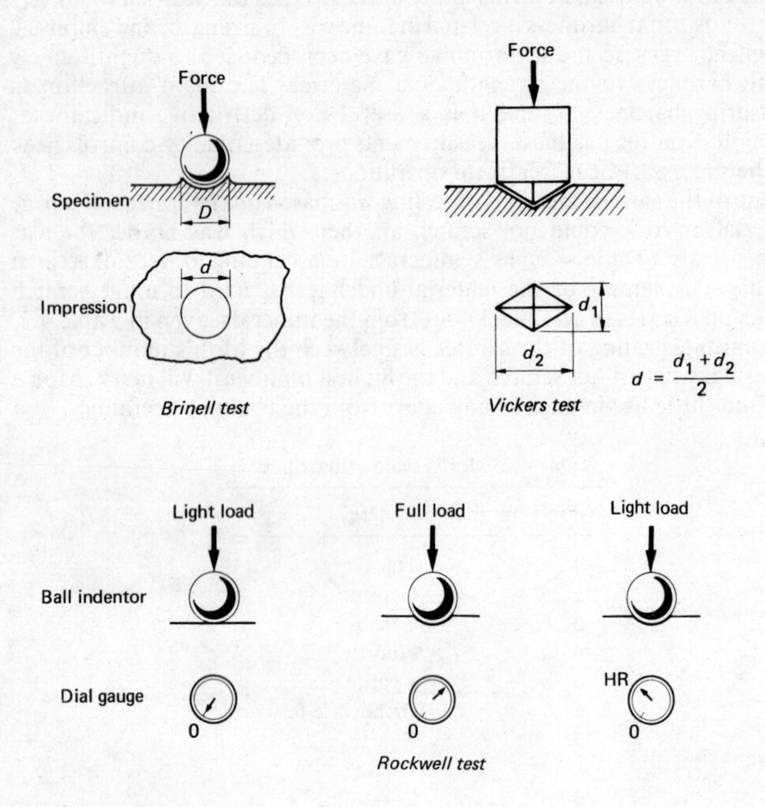

Fig. 4.13 Hardness testing

In the *Brinell test*, a 10 mm diameter hardened steel ball is used as an indentor. When the force has been removed, the diameter of the impression is measured with a calibrated microscope. The hardness number is given by

$$\text{Brinell hardness (HB)} = \frac{\text{applied force (kgf)}}{\text{spherical area of indentation (mm}^2\text{)}}$$

$$= \frac{F}{\frac{1}{2}\pi D[D - \sqrt{(D^2 - d^2)}]}$$

where D = diameter of ball

and d = diameter of impression

The *Vickers diamond* test is similar to the Brinell except that the indentor is a pyramid-shaped diamond. The angle between opposite faces of the pyramid is 136°. The impression is diamond-shaped, and both diagonals are measured to allow for any asymmetry which may result if the surface is not at right angles to the indentor.

$$\text{Vickers hardness (HV)} = \frac{\text{force applied (kgf)}}{\text{surface area of indentation}}$$

$$= \frac{2\,F \sin(\theta/2)}{d^2}$$

where θ = angle between faces of indentor = 136°

and d = average length of diagonals

In practice, there is no need to calculate each result; standard tables giving hardness values corresponding to indentation size for a number of different applied forces are available for both Brinell and Vickers tests.

Up to hardnesses of about 250 to 275, Brinell and Vickers numbers are more or less comparable. Above 275, the Brinell test tends to give softer results.

In the *Rockwell test*, the depth of indentation is measured directly, thus avoiding the need to use a microscope. This eliminates a time-consuming operation, and the tests are quicker than with the Brinell or Vickers systems. Rockwell hardness machines are frequently used on the shop floor for quality-control checks.

The machine is fitted with a dial gauge which indicates the depth to which the indentor has penetrated. The sequence of operation is:

a) a light load is applied to the indentor, to ensure good contact with the metal surface;
b) full load is applied, forcing the indentor into the metal;
c) full load is removed, leaving a light load to maintain contact;
d) the dial now indicates the depth of the indentation (D).

In practice, the dial is calibrated inversely so that hardness numbers can be read direct, avoiding the need to calculate them. Two scales are commonly used: Rockwell B has a ball indentor and a load of 100 kgf, while for Rockwell C a 150 kgf load is used with a diamond indentor. The scale used should be quoted with the hardness number.

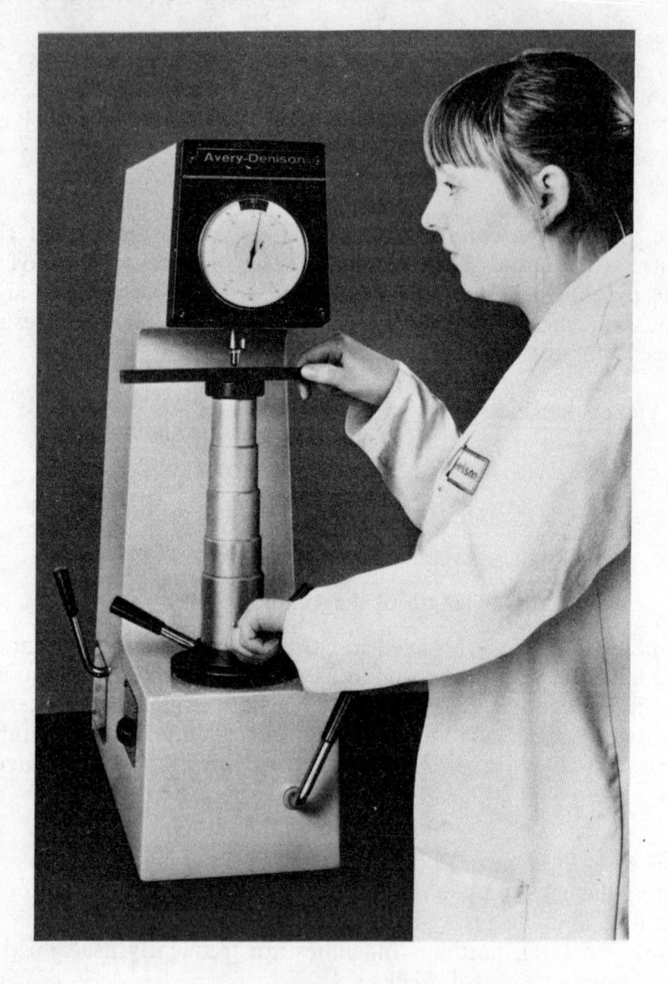

Fig. 4.14 Avery-Denison direct-reading Rockwell hardness-testing machine (model 6407)

5 Structure of polymers

Plastics have become so much a part of our daily lives that we rarely stop to consider the wide range of properties offered by readily available materials. The bottles used for washing-up liquid are flexible and do not split or tear easily. Plastics storage tanks are rigid and have good anti-corrosion properties. Fibre-reinforced plastics have high tensile strength and can be used in the fabrication of car bodies.

It is not easy to define a plastics material precisely. In some ways even glass can be considered as a plastics in that it is a non-metallic material which can be readily moulded into shape. For purposes of this chapter we shall define the plastics in which we are interested as materials which are made up of long-chain molecules based on carbon and hydrogen.

In chapter 2 we stated that the basic unit of a long-chain molecule is a *mer* and that a large number of mers can be joined together to form a chain. Polyethylene - or, as it is more commonly known, polythene - was given as a typical example of a plastics or polymer which could be produced by adding a number of mers together. There are, however, many other possible ways of building up a long-chain molecule. Each different combination gives rise to a characteristic set of properties, and in this chapter we will be examining the ways in which the structure of the molecule dictates the properties of the polymer.

5.1 Bonding in carbon compounds

When we look at the structure of organic compounds, we are principally concerned with covalent bonding between carbon and other elements such as hydrogen, oxygen, nitrogen, chlorine, and fluorine.

Carbon has a valency of four and can form strong covalent bonds. For example, two carbon atoms combine with each other and with six hydrogen atoms to form a gas called ethane. The structure of a molecule of this gas, which has the formula C_2H_6, is

```
    H   H
    |   |
H - C - C - H
    |   |
    H   H
```

With this arrangement, the four valencies are used up for each of the carbon atoms and there are single bonds between these and the hydrogen atoms.

The two carbon atoms could also be combined with only four hydrogen atoms as in ethylene, C_2H_4. If there are only two hydrogen atoms for each carbon atom, the latter must have spare bonds. In the molecule of ethylene, these are used to produce a double bond between the two carbon atoms:

```
H   H
|   |
C = C
|   |
H   H
```

A similar situation exists with acetylene, C_2H_2. There is now only one hydrogen atom, i.e. one single bond for each carbon atom. This means that there is a triple bond between the carbon atoms:

$$H-C\equiv C-H$$

It might be thought that double and triple bonds would be stronger than single bonds; however, the reverse is the truth. Only a single covalent bond is needed to hold the carbon atoms together, so, in a sense, the two extra bonds in acetylene are surplus to requirements and readily associate with other atoms. This is another way of saying that acetylene is highly reactive – a fact that is of great importance in the use of this gas as a source of heat for welding and cutting.

The three gases quoted above can therefore be arranged in order of increasing chemical reactivity:

```
                    H   H
                    |   |
ethane, C2H6    H - C - C - H      lowest reactivity
                    |   |
                    H   H

                    H   H
                    |   |
ethylene, C2H4      C = C
                    |   |
                    H   H

acetylene, C2H2   H - C ≡ C - H    highest reactivity
```

When all the bonds in an organic compound are single, it is said to be *saturated.* An *unsaturated* compound therefore contains one or more double or triple bonds.

Saturated compounds

The simplest saturated compounds are a group of hydrocarbons known as alkanes. These include many familiar substances such as the gases methane and propane and the liquids petrol and paraffin. (A hydrocarbon is a compound which contains only carbon and hydrogen.) All the compounds in this group have the same general molecular structure:

```
methane, CH4          H
                      |
                    H-C-H
                      |
                      H

ethane, C2H6          H H
                      | |
                    H-C-C-H
                      | |
                      H H

propane, C3H8         H H H
                      | | |
                    H-C-C-C-H
                      | | |
                      H H H
```

There are two hydrogen atoms for each carbon atom and, in addition, there are two hydrogen atoms to satisfy the spare bonds at each end of the chain.

We can write a general formula for the alkanes: $C_n H_{2n+2}$. The gases contain up to four carbon atoms; petrol usually has between five and nine carbon atoms; and the heavy oils can have twenty or more carbon atoms. The alkane gases are important not only for use as fuel but also as a basis for the manufacture of many polymers.

5.2 Monomers and polymers

If we take another look at the structure of ethane, we can see that it is made up of two mers plus two extra hydrogen atoms:

```
   ¦ H ¦ H ¦
   ¦ | ¦ | ¦
 H-¦-C-¦-C-¦-H
   ¦ | ¦ | ¦
   ¦ H ¦ H ¦
    mer  mer
```

Suppose the two hydrogen atoms were taken away:

```
H H
| |
C=C
| |
H H
```

the compound becomes unsaturated and we arrive at ethylene.

An important feature of ethylene is its ability to combine with itself. Two or more molecules of ethylene can be linked together to form a longer chain. Compounds which, like ethylene, can be combined in this way are called *monomers* and are the basic unit in the formation of many useful plastics. A *polymer* is a molecule produced by linking together a large number of monomers. The same monomer may be used throughout the chain, or the molecule may be constructed from two monomers arranged in regular or random order. Usually between 1000 and 100 000 monomers are present in a polymer molecule.

Typical monomers

Variations in the composition of monomers can be produced by replacing one or more of the hydrogen atoms in ethylene. Some of the simplest monomers used in the production of polymeric materials are

ethylene, C_2H_4

```
H  H
|  |
C=C
|  |
H  H
```

vinyl chloride, C_2H_3Cl

```
H  H
|  |
C=C
|  |
H  Cl
```

vinylidene chloride, $C_2H_2Cl_2$

```
H  Cl
|  |
C=C
|  |
H  Cl
```

Other monomers are more complex and involve the replacement of one hydrogen atom by a molecule of another organic compound.

Vinyl acetate incorporates an acetate configuration consisting of two carbon, three hydrogen, and two oxygen atoms:

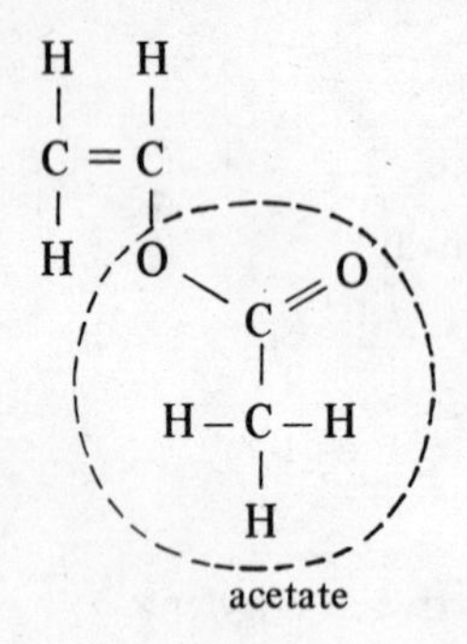

In *isopropylene*, the replacement consists of one carbon and three hydrogen atoms:

```
H     H
|     |
C ═══ C
|     |
H  H–C–H
      |
      H
```

Greater complexity is found in *styrene*, which incorporates a *benzene ring*. In the ring, six carbon atoms are joined together, with single and double bonds alternating round the ring. Five of the carbon atoms are also bonded to a hydrogen atom, while the remaining one links to a carbon atom in the ethylene monomer:

```
      H   H
      |   |
      C = C
      |   |
      H   C
        //  \
   H–C        C–H
     |        ||
   H–C        C–H
       \\   /
          C
          |
          H
      benzene
```

5.3 Polymerisation

The chemical reaction by which monomers are combined together to produce polymers is called *polymerisation*. In many cases the name of the polymer is very simply derived from that of the monomer:

Monomer	*Polymer*
ethylene	polyethylene or polythene
vinyl chloride	polyvinylchloride (PVC)
isopropylene	polypropylene
vinyl acetate	polyvinyl acetate (PVA)
styrene	polystyrene

The method used to bring about the combination depends on the monomer(s) involved. Broadly speaking, there are three main types of polymerisation: addition polymerisation, copolymerisation, and condensation polymerisation.

Addition polymerisation As the name implies, an addition polymer is produced by adding monomers together. For example, adding ethylene monomers to give polythene:

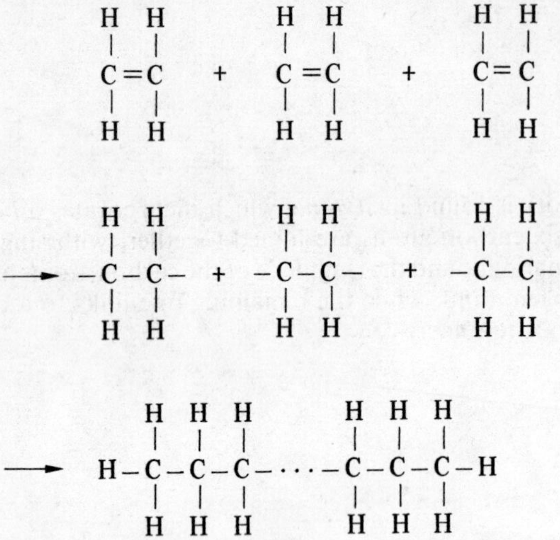

```
   H  H         H  H          H  H
   |  |         |  |          |  |
   C=C    +     C=C     +     C=C
   |  |         |  |          |  |
   H  H         H  H          H  H

   H  H         H  H          H  H
   |  |         |  |          |  |
—→ –C–C–   +   –C–C–    +    –C–C–
   |  |         |  |          |  |
   H  H         H  H          H  H

      H  H  H          H  H  H
      |  |  |          |  |  |
—→  H–C–C–C–·····–C–C–C–H
      |  |  |          |  |  |
      H  H  H          H  H  H
```

Another example of addition polymerisation is provided by polystyrene:

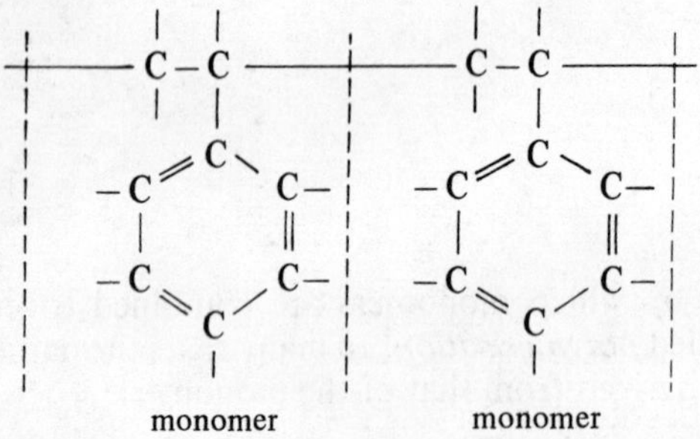

```
|     |  |          |      |  |          |
+-----C–C-----------+------C–C-----------+
|     |  |          |      |  |          |
|        C          |         C          |
|    –C//  \  C–    |    –C//  \  C–     |
|     |       ||    |     |       ||     |
|    –C        C–   |    –C        C–    |
|      \\    /      |      \\    /       |
|         C         |         C          |
|         |         |         |          |
      monomer             monomer
```

Note: the hydrogen symbols have been omitted for clarity.

Copolymerisation Some monomers do not combine with themselves but can be linked with monomers of a different composition. It may also be desirable to combine two monomers to achieve specific properties. This process is known as copolymerisation. The properties of the polymers so produced are different from those of the monomers involved. Typical combinations are ethylene + vinyl acetate and vinylidene chloride + vinyl chloride. The structure of the chain is in the form

$$\cdots - \Delta - \bigcirc - \bigcirc - \Delta - \bigcirc - \bigcirc - \bigcirc - \Delta - \bigcirc - \cdots$$

$\longleftarrow$ A $\longrightarrow$

where Δ = ethylene monomer

and $\bigcirc$ = vinyl-acetate monomer

In detail, A would have the structure

```
 H  H¦ H  H    ¦ H  H    ¦ H  H ¦
 |  |¦ |  |    ¦ |  |    ¦ |  | ¦
-C--C+-C--C----+-C--C----+-C--C-¦
 |  |¦ |  |    ¦ |  |    ¦ |  | ¦
 H  H¦ H  O   O\H   O   O\H   H ¦
     ¦     \ //  \   \ //  \    ¦
     ¦      C     ¦    C    ¦   ¦
     ¦      |     ¦    |    ¦   ¦
     ¦   H--C--H  ¦ H--C--H ¦   ¦
     ¦      |     ¦    |    ¦   ¦
     ¦      H     ¦    H    ¦   ¦

ethylene  vinyl acetate   vinyl    ethylene
                          acetate
```

Condensation polymerisation In this type of polymerisation, two relatively simple long-chain molecules enter into a chemical reaction which forms a more complex molecule with water as a by-product. Most of the hard plastics, such as Bakelite and melamine, are produced by condensation polymerisation.

5.4 Structure of a polymer molecule

So far we have assumed that monomers link together in a straight line and we have depicted the chain thus:

$$-C-C-C-C-C-C-$$

Molecules which have this structure are called linear.

This is a very convenient way to draw the molecular structure on a page of a book, but in reality we are dealing with a three-dimensional system. The four covalent bonds of a carbon atom are identical and are symmetrically disposed in space. It is difficult to show this with a two-dimensional drawing. We need to visualise the carbon atom located at the centre of a tetrahedron – this is a pyramid with a triangular base, each face of which is an equilateral triangle. The bonds of the carbon atoms point outwards to the corners of the tetrahedron (fig. 5.1). The angle between any two adjacent bonds is 109.5°.

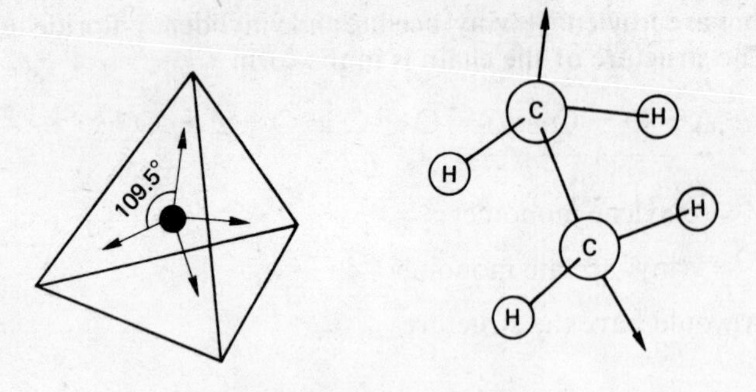

Fig. 5.1 Bond angles for a carbon atom

In a simple arrangement, the bonds between carbon atoms in a polymer chain go 'up' and 'down' alternately:

C C C
C C C C

(Remember that we are trying to show in two dimensions what happens in a three-dimensional model.)

This regular pattern need not occur. To take an over-simplified view, as the bonds are symmetrical and identical, the one which is chosen by the next carbon atom in line is not necessarily influenced by the orientation of the preceding atom. In other words, because the previous bond went 'up', it does not follow that the next must go 'down' – it can go 'up'. There is, therefore, a high probability that the chain may snake along its length (fig. 5.2):

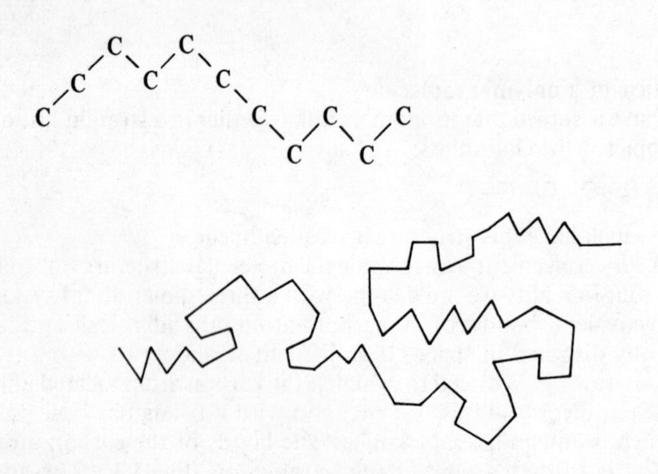

Fig. 5.2 Zig-zag chain in a linear molecule

If the chain has a non-symmetrical distribution of atoms, e.g. in the case of rubber

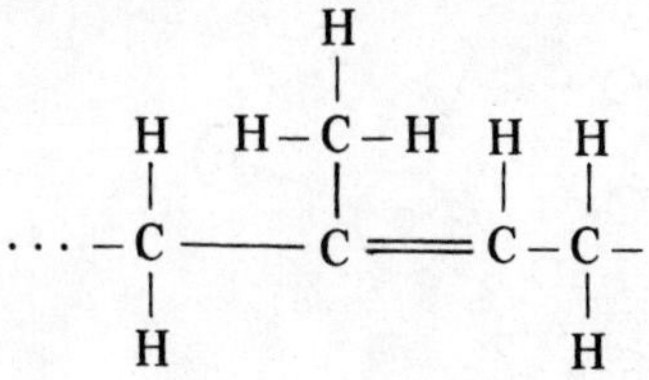

```
             H
             |
    H    H - C - H   H   H
    |        |       |   |
··· - C ———— C ═══ C - C -
    |                    |
    H                    H
```

it will take up the configuration of a helical coil:

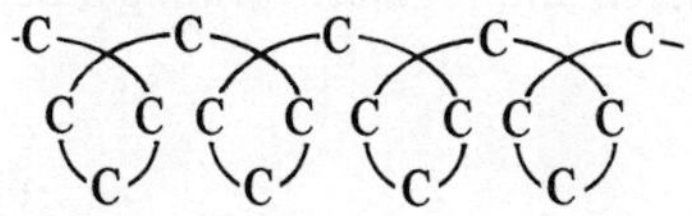

```
-C-⁀-C-⁀-C-⁀-C-⁀-C-
C  C C  C C  C C  C
 \C/  \C/  \C/  \C/
```

Effect of stress on linear molecules

When a force is applied to a polymer molecule, it tries to straighten out the chain. Some of the large elongations noted when testing polymers are attributable to this behaviour. Molecules which lie parallel to each other can also slide, giving greater elongation. The zig-zag arrangement offers an appreciable extension as the bonds are strained into the line of the applied force. The helical coils of the rubber molecule straighten out under stress, giving very large extensions. This explains the high percentage elongation which is a notable characteristic of rubber.

Materials which contain long coiled molecules are called *elastomers*.

Branching and cross-linking

Some monomers can be persuaded to form a branch from the linear chain:

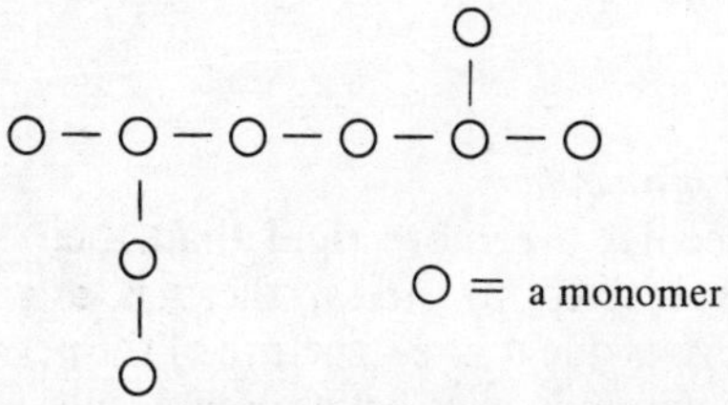

```
                O
                |
O - O - O - O - O - O
    |
    O           O = a monomer
    |
    O
```

These branches can often be cross-linked to neighbouring chains by a chemical reaction called *curing*:

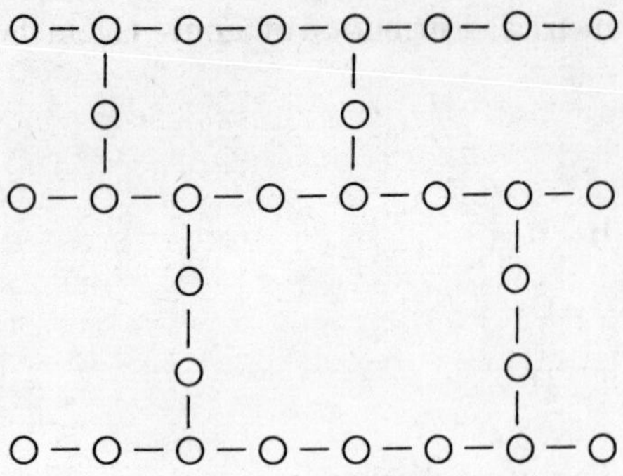

```
O—O—O—O—O—O—O—O
  |       |
  O       O
  |       |
O—O—O—O—O—O—O—O
    |       |
    O       O
    |       |
    O       O
    |       |
O—O—O—O—O—O—O—O
```

The cross-linking may also be ordered and regular, forming a definite pattern or network:

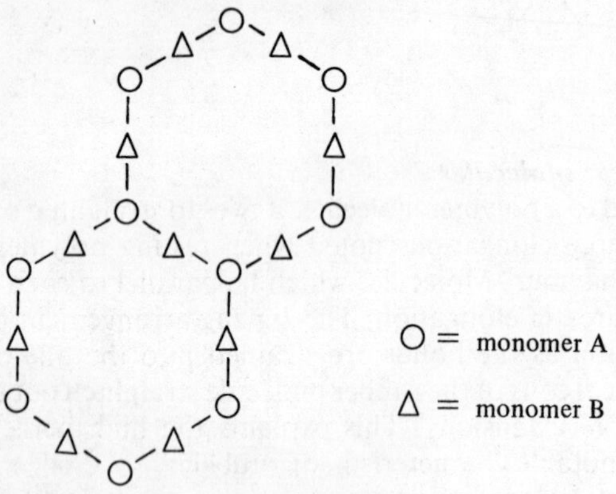

```
O = monomer A

Δ = monomer B
```

These networks should not be confused with the benzene ring – O and Δ are two different complete monomers. It is feasible, though, that one of the monomers in a network could itself contain a benzene ring.

More often than not, branching and cross-linking are irregular and do not give an ordered pattern. Frequently the molecule gives the appearance of tangled wool.

Effect of stress on a cross-linked molecule

Branched and cross-linked molecules are more rigid than their linear counterparts. When they are subjected to stress, there is a greater resistance to movement. In part this is due to the structure of the network, but there is also a greater risk of entanglement between molecules. With linear molecules, where the stress tends to straighten the chain, there is a good chance that they can move freely relative to each other. With a branched structure, one molecule trying to move past another could meet with opposition due to branches locking together. Similarly, complex cross-linked molecules can interact.

In general, therefore, branched and cross-linked molecules give hard, strong, but sometimes brittle materials. Rubber is normally a very pliable material with enormous elasticity. When it is vulcanised, however, it becomes harder and loses most of this elasticity. In the vulcanisation process, sulphur atoms are introduced to form cross-links between neighbouring rubber molecules (fig. 5.3). These effectively prevent the helical coils from stretching under the effect of applied stress.

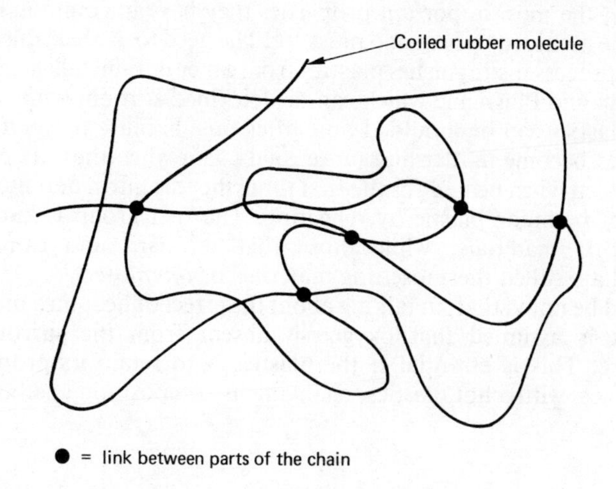

Fig. 5.3 Cross-linking in lightly vulcanised rubber

Bakelite is a good example of a very hard plastics which derives its hardness from a rigid three-dimensional network.

Plasticisers

From what has been said so far, it is clear that the ductility of a polymer depends on the ease with which the molecules can move past each other. In some polymers, ductility is improved by the use of materials called *plasticisers*. When these are added to a polymer, they go between neighbouring molecules, increasing the distance between them. The intermolecular forces are thus reduced, and it is easier for the chains to slide past each other when subjected to stress.

The effect of a plasticiser can be observed with polyvinylchloride (PVC). In the unplasticised condition, PVC is hard and rigid and it is used for the manufacture of gutters and pipes used in building. By contrast, the addition of a plasticiser enables PVC to be made into thin very pliable film used for packaging and rainwear.

6 Plastics in practice

6.1 Thermoplastics and thermosets

When we start to consider the use of plastics in manufacture, we realise that one of the most important properties they have in common is their ability to be moulded by heat and pressure. The need to use heat during the moulding process distinguishes plastics from other non-metallic materials such as wax and Plasticine which can be deformed at room temperature.

Some plastics can be moulded time after time because they soften on heating and become harder and more rigid again when they are cooled. Others harden when heated for the first time: they set into a definite shape and cannot be made plastic by reheating. The first group is known as *thermoplastic* materials, while those that set hard and cannot be resoftened are called thermosetting materials or *thermosets*.

It should be noted that, in talking about the effect of heat on a moulded sample, it is assumed that oxygen is absent from the surrounding atmosphere. This is essential if the plastics is to retain its properties. Oxygen reacts with a hot plastics, changing its composition and hence its properties.

Thermoplastic materials These contain linear or branched long-chain molecules which are not interconnected. They have high inherent plasticity which increases as the temperature is raised. They do not undergo chemical changes when heated and, on cooling, their plasticity is retained because the structure is unchanged.

The plastic behaviour of a particular thermoplastic material – especially when subjected to heat – is, of course, dependent on the composition and therefore the structure. It is also influenced by the presence of plasticisers.

Some common thermoplastics are

polyethylene (polythene),
polyvinyl chloride (PVC),
polystyrene,
polypropylene,
nylon.

Thermosetting materials These usually have a cross-linked network structure which gives them their characteristic hardness and rigidity. This structure is established during the moulding operation. If thermosets are reheated, the network remains intact until a temperature is reached when the plastics disintegrates. This is why thermosets cannot lose their rigidity and become plastic.

Some common thermosets are

phenol formaldehyde (Bakelite),
urea formaldehyde,
melamine formaldehyde,
polyester resin,
epoxy resin.

6.2 Moulding plastics

Most plastics items are made by moulding. The product may be the finished shape or may need further forming.

Moulding thermoplastics

Two main techniques are used to mould thermoplastics: extrusion and calendering.

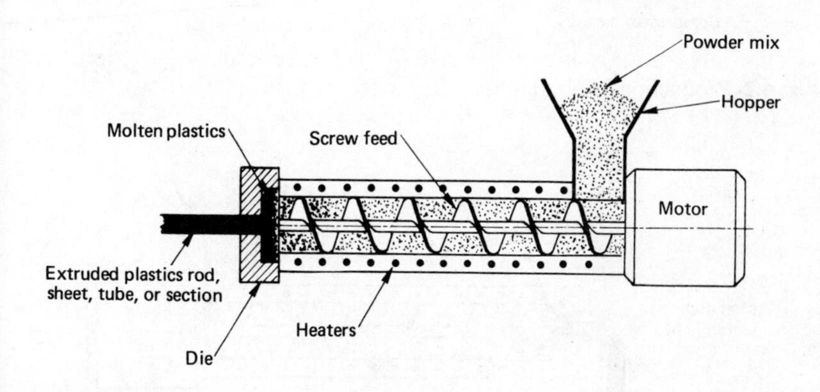

Fig. 6.1 Continuous extrusion of thermoplastics

In *continuous extrusion* (fig. 6.1), the powdered polymer or monomer is fed by a screw along a cylindrical chamber. As the powder moves toward the die, it is heated and melts. The molten plastics is forced through the die, where it is formed into the desired shape. The principle of extrusion is incorporated into a large number of manufacturing sequences. In its simplest form it can produce rod, tube, section, or sheet. The die can also be shaped to produce a tube into which compressed air or nitrogen is blown to give a film bubble from which sheet as thin as 0.7 mm can be obtained (fig. 6.2).

A special application of extrusion is *injection moulding* (fig. 6.3). The principles are similar to those of extrusion, but the process is operated on a

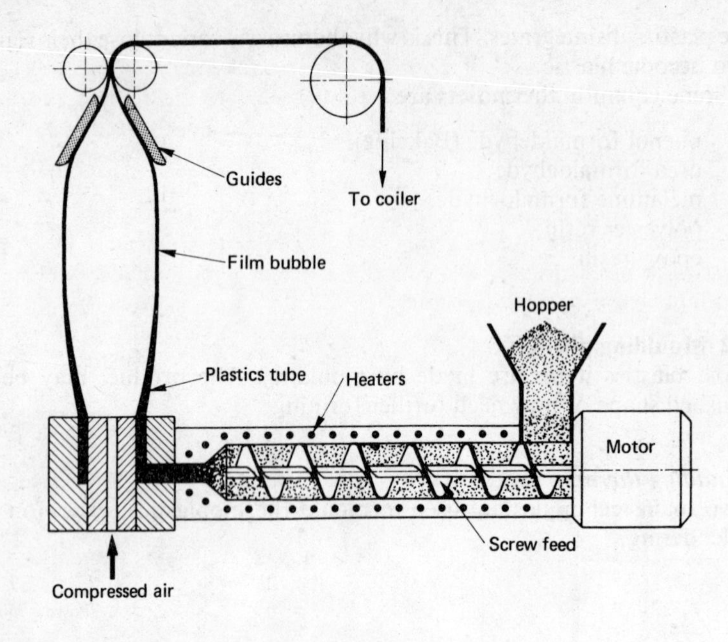

Fig. 6.2 Production of plastics film

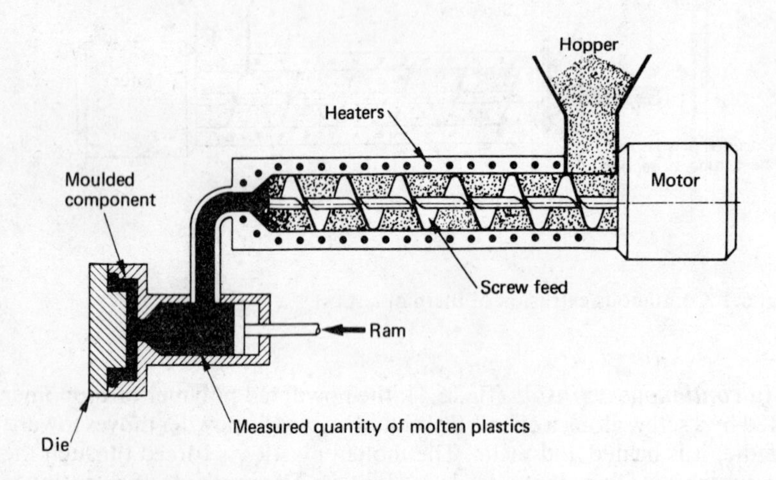

Fig. 6.3 injection moulding of thermoplastics

unit basis. A measured amount of powder is introduced into the heated chamber, where it is melted. The molten plastics is then forced into a die containing a cavity which has the shape of the component being moulded. The plastics cools under pressure, to avoid 'shrink' marks on the surface.

In *blow moulding* (fig. 6.4), the plastics is extruded as a tube. At measured intervals, a split mould is closed around the tube, sealing off the lower end. Air is blown into the tube, inflating it to the shape of the mould.

The powder used in the extrusion press can be a monomer which is polymerised in the heated sections of the chamber. However, it is not always easy to control the polymerisation reaction, especially when extruding thick sections, and it is often preferable to produce the polymer in a separate operation. It is then ground into a powder for feeding into the extrusion press. Plasticisers can be mixed with the polymer powder, for example when making thin film.

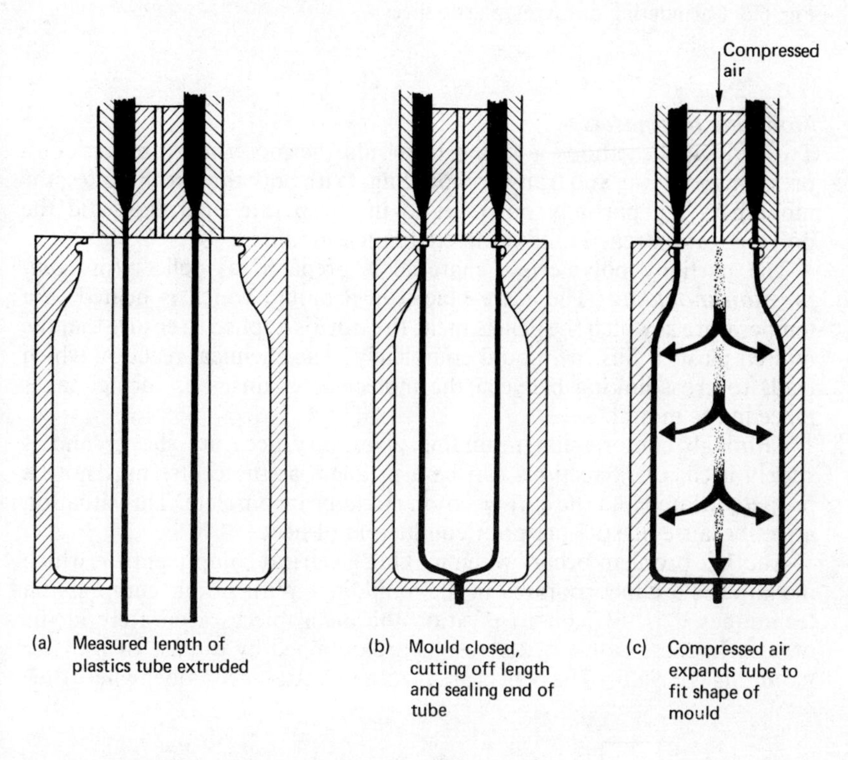

Fig. 6.4 Blow moulding of thermoplastic bottles

Sheets, especially in PVC, can also be produced by *calendering* (fig. 6.5). The polymer is first mixed with plasticisers and other additives such as colouring agents. The mix is then heated for a short time to produce a rough sheet which is fed through a series of rolls. These gradually reduce the thickness and, at the same time, give a smooth or textured surface as required. The finished thickness is determined by the setting of the gap between the last pair of rolls.

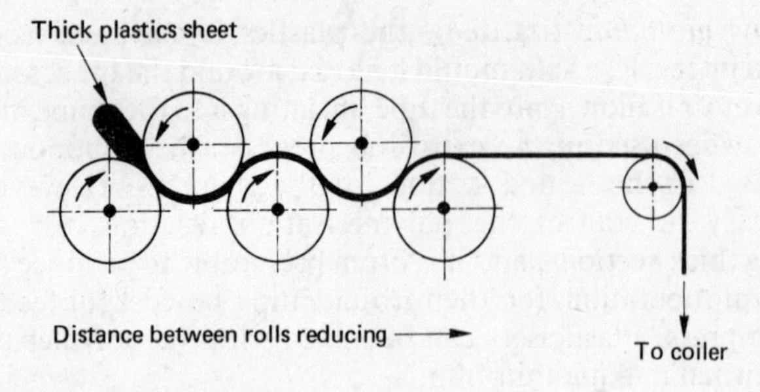

Fig. 6.5 Calendering of thermoplastic sheet

Moulding thermosets

Two principal methods are used to mould thermosetting plastics: compression moulding and transfer moulding. With both these techniques, the monomers are partially polymerised in a separate operation and the polymerisation reaction is completed in the mould.

The partially polymerised material is prepared as pellets for *compression moulding*. These are placed in a mould which is heated to a temperature at which the pellets melt. Pressure is applied to ensure that the molten plastics fills the mould completely. The chemical reaction which leads to cross-linking between the molecular chains, i.e. curing, takes place in the mould.

Although compression moulding gives very accurate shapes and is widely used, thick sections can be a problem as the centre may not be properly heated and the curing action might be incomplete. This situation arises because plastics are poor conductors of heat.

Another problem occurs when making electrical components in which metal parts are incorporated in the moulding. With simple compression techniques it is difficult to position the metal pieces accurately in the mould. In these cases, better results are obtained by the use of *transfer moulding* (fig. 6.6). The pellets are placed in an outer chamber where they

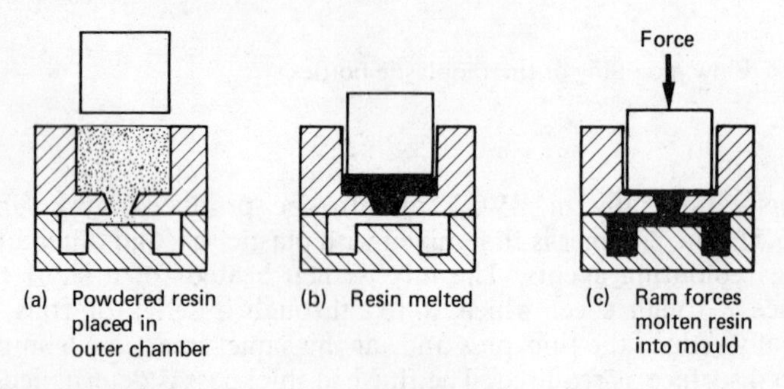

Fig. 6.6 Transfer moulding of thermosetting plastics

are melted. The molten plastics is then forced into the heated mould. Great care is taken to ensure that the curing reaction takes place after the liquid has completely filled the mould. In some ways, transfer moulding is similar to the injection-moulding process used for thermoplastics.

6.3 Typical thermoplastics

The range of thermoplastics is continually increasing as chemists discover new combinations of organic compounds. Even within one type of polymer there are different grades resulting from variations in the length and configuration of the molecular chains. The scope of thermoplastics can be appreciated by considering five of the most common types.

Polyethylene

Polyethylene is also more commonly known as polythene. It is the product of the polymerisation of ethylene. The molecule is a simple linear chain of carbon atoms each attached to two hydrogen atoms (see pages 17 and 23).

There are two types of polyethylene. *High-density* polythene is made by polymerising liquid ethylene at low pressures. If ethylene vapour is polymerised at high pressures, *low-density* polythene is produced.

The properties of polythene are closely related to its density. The low-density material is easily moulded, has good impact properties, and is flexible down to low temperatures (about −60°C). As the density is increased, the tensile strength and hardness increase but the impact strength falls – i.e. the material becomes more brittle. The secant modulus (page 47) also increases, which means that the polythene becomes more rigid.

Polythene is widely used for packaging, because it can be readily produced as a thin film. It has good resistance to chemical attack and is suitable for both industrial and domestic pipework. It can also be used as an insulator in electrical systems.

Polyvinylchloride (PVC)

PVC is another thermoplastic which is widely known through its domestic everyday applications. It is made by the polymerisation of vinyl chloride.

The molecules are linear, but there are appreciable attractive forces between them due to the presence of chlorine atoms. The forces limit the freedom of movement of the molecules, and the polymer has appreciable rigidity. Greater flexibility and softness can be achieved by the addition of plasticisers (see page 67). The amount of plasticiser added is adjusted to give the required properties and can be between 5% and 50%. The plasticity can also be controlled by copolymerisation with vinyl acetate, which acts as an internal plasticiser. The majority of PVC components are manufactured from material which contains a plasticiser.

PVC tends to decompose when subjected to strong sunlight. This tendency is controlled by the addition of a *stabiliser*. Lead compounds are used to stabilise PVC.

Like polyethylene, PVC has good resistance to chemical attack and is a useful electrical insulator. It is considerably more flame-resistant than polythene.

PVC is used as an insulator for electrical wiring, as a lining for chemical tanks and pipework, and in the manufacture of flexible and rigid tubing, hoses, rainwear, protective clothing, and a host of other domestic goods.

Polypropylene

The monomer propylene can be polymerised to give a linear chain with the CH_3 groupings on one side of the molecule:

```
      H   H        H   H        H   H   H
      |   |        |   |        |   |   |
  ----C - C -------C - C -------C - C - C----
      |   |        |   |        |   |   |
      H   |        H   |        H   |   H
          |            |            |
      H - C - H    H - C - H    H - C - H
          |            |            |
          H            H            H
```

This gives a stronger more rigid material than polyethylene.

An important characteristic of polypropylene is its fatigue strength when subjected to repeated bending. This means that a strip of polypropylene can be flexed very many times without cracking. It is thus suitable for use as a hinge which can be injection moulded as an integral part of a component.

Filaments or threads made from polypropylene are woven into lightweight ropes. These are waterproof, do not rot, and float on water – the polymer's relative density is only 0.9. Polypropylene's good resistance to flexing ensures that the ropes do not break after repeated coiling.

Polypropylene has a relatively high softening point (about 140°C) and can be used for bottles which must contain boiling water or be sterilised by steam. It also has good resistance to impact and is used for battery cases and bottle crates.

Polystyrene

Styrene is a complex monomer containing a benzene ring, but it polymerises to give a linear molecule (page 61). In the liquid form, polystyrene has good fluidity and is ideally suited to injection moulding. It is probably best known for its use in model kits, where the ease of moulding enables very fine detail to be produced in the moulded parts.

Although polystyrene is a good insulator, its high dielectric strength and very low loss factor are more often the reasons for its use in electrical components.

By itself, the polymer is hard and brittle but it can be toughened by blending with synthetic rubber. This toughened material is used to make containers which have self-sealing lids. It is resistant to attack by strong acids and alkalis, but it is soluble in petrol.

Polystyrene can also be produced in an expanded form. This has a cellular structure, is very light, and has excellent thermal-insulating characteristics. Expanded polystyrene is widely used in buildings and also provides a good packaging material which resists shock.

Nylon

Nylon is a name given to a whole group of complex thermoplastics (polyamides) formed by reactions between an organic acid and an organic compound (an amine). The molecules are linear, and the properties depend on the compounds from which the polyamide has been made. The various nylons are identified by an index number which indicates how many carbon atoms were in the compounds which took part in the reaction. Nylon 6,6 is made from two organic compounds each containing six carbon atoms in their molecules. Nylon 6,10 starts from one compound with six atoms and one with ten, and so on. Where only one chemical is involved in the polymerisation reaction, there is a single index number, e.g. nylon 6 or nylon 11.

Almost everyone must be familiar with nylon as a filament or thread used to produce fabric. As a bulk material, however, nylon is strong, has good resistance to abrasion and impact, and does not soften at temperatures below 200°C. It can be readily machined and is used for bearings, gears, and cams, as it does not require lubrication.

6.4 Thermosetting materials

Earlier in this chapter we saw that thermosets derive their properties from a rigid network structure which does not allow neighbouring molecules to slide past each other. Once this network has been established, the polymer cannot be softened but decomposes if it is heated to a high temperature. This means that production of a component from a thermoset requires firstly the formation of partially polymerised material. If the polymerisation reaction is prevented from going to completion, a network is not formed and only linear molecules exist.

Thermosets are produced by reacting two organic compounds in a steam-heated 'kettle'. The organic compounds are dissolved in water, and a catalyst may be needed to bring about the reaction. (A catalyst is a chemical which affects the rate at which the compounds react with each other but remains unchanged at the end of the reaction.) The products of the reaction are a resin and water. These evaporate from the boiling liquid in the kettle, and the resin is separated from the water by condensation. The reaction is controlled so that the resin which is collected contains only linear molecules.

When the resin is reheated during moulding, the polymerisation process is restarted and links are set up between the molecules. Usually a curing agent is added to ensure that the reaction goes to completion.

The curing process is accompanied by a contraction in volume. This shrinkage is offset by adding filler to the powder used for the moulding operation. The filler must be inert; in other words, it must not interfere with the curing reaction. Fillers reduce the cost of the component by replacing some of the expensive resin. They can also be used to confer specific properties.

Fillers generally make up 50% to 75% of the volume of the powder used for moulding. Some typical fillers are listed in Table 6.1.

Table 6.1 Typical fillers for thermosets

Filler	Reason for addition
Wood flour	Reduces amount of resin required and improves mechanical properties.
Fabric, nylon, and paper pulp	Improve impact strength.
Asbestos	Gives better resistance to heat and acid attack.
Mica	Increases electrical resistance.
Graphite	Lowers surface friction.

Thermosets are widely used for mouldings in electrical equipment (plugs, sockets, casings, and insulating bushes and sleeves), engineering equipment (handles, cases, screw caps, and brackets), and domestic ware (kitchen equipment, food mixers, tableware, and door fittings).

Another important application of thermosetting materials is in the production of laminates. These are made by bonding together layers of paper or fabric with the thermoset. Cotton, nylon, and woven glass-fibre sheeting are very often used for this purpose. The laminate can be made either in the form of a sheet or as a bar or rod which can be machined into shape. Laminates are widely used in the manufacture of bearings and gear wheels.

The four most common groups of thermosets are:

a) phenolic resins,
b) amino resins,
c) polyester resins,
d) epoxy resins.

Within each group there are a number of variations.

Phenolic resins (PF resins)

These are produced by a condensation reaction between phenol and formaldehyde. They are resistant to attack by the common acids and solvents but are decomposed by hot alkalis. Unfortunately, they absorb small amounts of water, which can cause them to become slightly swollen. PF resins have good resistance to heat, hence their use in the manufacture of ash trays. They are naturally dark, and black or brown pigments are often added to give a uniform colour. Probably the best-known phenol-formaldehyde resin is Bakelite.

Amino resins

Amines are a group of organic compounds which contain nitrogen. When they react with formaldehyde, thermosetting resins are condensed out. The two most commonly used resins in this group are urea formaldehyde (UF resin) and melamine formaldehyde (MF resin).

UF resins have similar properties to PF resins but have better electrical properties. They also have the advantage of being colourless, and a range of pigments can be used for decorative effect.

MF resins have better resistance to heat than UF resins, which tend to crack on prolonged heating. They are therefore preferred for electrical components which operate in hot conditions. A major application of MF resins is the manufacture of domestic tableware. They have no taste and are not attacked by tea, coffee, or fruit juices.

Polyester resins

If an organic acid is reacted with alcohol, the products are an ester and water. There are a host of esters which can be polymerised to give polyester resins. The curing temperature of the resins in this group is controlled by the composition. Some polyesters will cure without the application of heat; while, with others, cross-linking will not take place until the temperature has been raised to about 100°C. The time required for curing is also dependent on the temperature. Without heating, curing can take many hours, while at elevated temperatures the reaction can be completed in a matter of minutes.

The principal uses of polyester resins are the manufacture of glass-fibre-reinforced components (car bodies, machine casings and covers, and storage boxes), filaments which are woven into fabric, and heat-resistant lacquers for surface finishes on furniture.

Epoxy resins

As with polyesters, there are a number of different epoxy resins. They are extremely tough, are resistant to heat, and are not affected by the majority of acids and alkalis. They differ from many of the thermosetting materials in that they are normally cured without heating, by the addition of a hardener. This is an advantage in electrical equipment, where delicate components can be encapsulated in a protective block of resin without being subjected to the risk of damage by heat.

A very important application of epoxy resins is in jointing. They will adhere to the surface of many materials, including plastics, wood, ceramics, glass, and metals. Very strong lap joints can be made between both similar and different materials. This is particularly useful in situations where, for example, a glass-to-metal joint is required which would be difficult to achieve with other techniques.

Epoxy materials are usually supplied as two components: a resin and a hardener which are mixed just before use. The ratio of resin to hardener determines the time taken to cure.

7 Structure of metals

One important feature which distinguishes a metal from most other materials is that in the solid state it is crystalline. This is another way of saying that its atoms are arranged in a regular three-dimensional pattern which has long-range order (see chapter 2). Why should this be significant? Most of the properties which interest us in the use of metals in engineering are in some way related to the atomic arrangement. The properties obtained when metals are mixed together (i.e. in alloying) are dictated by the atomic structure. This also controls the way in which metals can be processed. Heat treatment and forming by plastic deformation are two examples of common metal-processing operations. How they can be used and the results achieved depend on the structure of the metal. If we are to understand how metals behave, we must be able to recognise the different arrangements of atoms which exist in commonly used metals.

7.1 Types of lattice structure

When discussing lattice structures, it is very convenient to think of atoms as solid spheres. We know from chapter 2 that this is not correct, of course; but, as we are interested only in the space occupied in the lattice by each atom, we do not need to consider its internal structure. This means that we can use spheres of, say, polystyrene or cotton wool to produce models which illustrate the various possible geometrical arrangements.

If we place a large number of spheres on a tray and line them up in rows, it can be seen that there are two basic arrangements. In the first (fig. 7.1(a)), the spheres are located at the corners of imaginary squares. Each sphere is in touch with four other spheres, and the tangents at points of contact are at right angles to each other. This is, therefore, a *square* arrangement. Alternatively, the rows can be displaced so that the spheres nestle between each other (fig. 7.1(b)). Each sphere is now in contact with six others which form a hexagon around it. It will be noticed that there is less space between the spheres with this arrangement, and for this reason it is often referred to as *close-packed*.

The next step in our investigation is to stack further layers of spheres on top of the first layer. With the square arrangement, we could position the spheres in the second layer immediately over those in the first, and so on with the third, fourth, and subsequent layers (fig. 7.2(a)). The result would be that the spheres would be sited at the intersections of a cubic space lattice (fig. 7.2(b)). It would also be possible to visualise the same arrangement being made up of a number of cubes stacked together. It

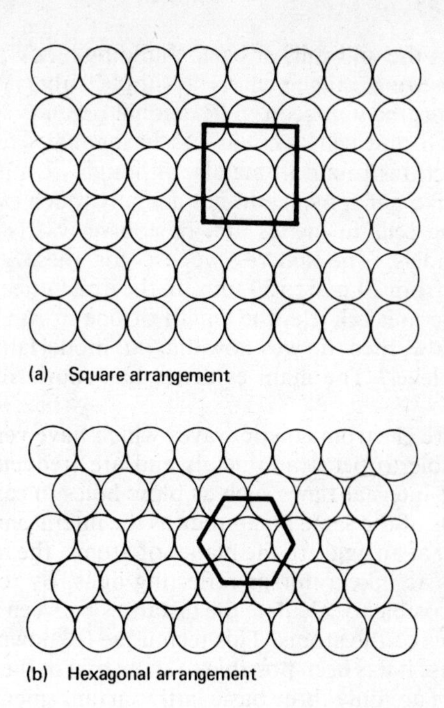

Fig. 7.1 Two basic arrangements for spheres

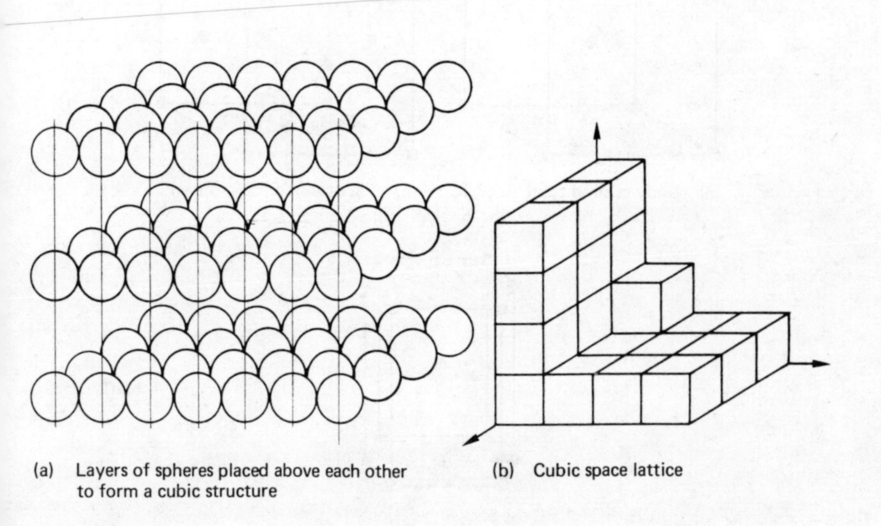

Fig. 7.2 Three-dimensional cubic arrangement of spheres

follows that the unit cell, or basic 'building block', for a lattice made up from a square arrangement is a simple cube. With a closed-packed arrangement, the unit cell is a hexagonal prism.

These two unit cells – cubic and hexagonal – feature in most of the lattice structures found in metals, although we must remember that the atoms, unlike our spheres, do not actually touch each other. The dimensions of the cells in metals are, of course, vastly different from those of our models. The spheres we use for these give cubes with sides measuring from 20 mm to 50 mm. In the iron lattice, the lattice parameter of the cubic unit cell (i.e. the length of one of its sides) is about 2.86×10^{-10} m. How, then, do we know that our model lattices are reproduced at an atomic level? The main evidence has come from X-ray diffraction studies.

X-rays are electromagnetic waves which have very short wavelengths. They are able to penetrate metals and are frequently used to show the presence of internal flaws such as blow-holes in castings. If X-rays with wavelengths of the same magnitude as the interatomic spacings in a metal are directed at an angle to the planes of atoms, the rays are reflected. The planes thus act like a mirror reflecting light. By recording the reflected rays, it is possible to calculate the distances between the planes and to plot the positions of the atoms. This technique is known as *X-ray diffraction* and by its use it has been possible to show that in the majority of common metals there are only three basic lattice arrangements (fig. 7.3):

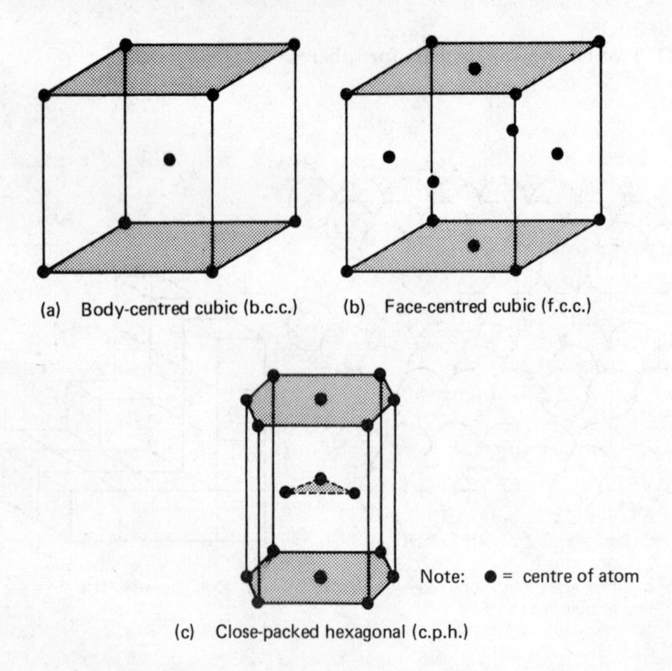

Fig. 7.3 Space lattices for a metal

a) body-centred cubic (b.c.c.),
b) face-centred cubic (f.c.c.), and
c) close-packed hexagonal (c.p.h.).

Body-centred cubic lattices have a cubic unit cell in which there is an atom at each corner and one in the centre of the body of the cube. Although a b.c.c. lattice appears to be simple, it is not very often found in commercially available metals. Iron is the best-known metal which has a b.c.c. lattice.

Face-centred cubic lattices also have an atom at the corner of each cube, but these are further apart than in the b.c.c. type. This leaves room for an additional atom at the centre of each face of the cube. Many metals have f.c.c. lattices; examples are aluminium, copper, and nickel. In general, the most ductile metals have a f.c.c. structure.

Close-packed hexagonal lattices are significantly different from the cubic structures. The unit cell is not symmetrical. There are three layers in each cell. The top and bottom layers consist of six atoms in a hexagon with one atom at the centre. The middle layer has three atoms in the form of a triangle. The vertical faces of the cell are rectangular. This is the least common of the lattice structures. Examples of metals with a c.p.h. lattice are zinc and magnesium - it is worth noting that these are among the least ductile metals.

Imperfections in crystal lattices

At first sight, there is no reason to suppose that the lattice structure should not be uniform throughout its length, breadth, and depth. The positions which atoms occupy are determined by the forces acting on them. Since all the atoms in a metal are essentially uniform, it is reasonable to expect that interatomic forces are the same at all points in the lattice. In practice, however, imperfections do occur. This may be because there was some interruption in the growth of the crystal from molten metal. Imperfections can also be created by the inclusion of an atom of another metal, whether added intentionally or present as an impurity. They are worthy of note because they can have a significant effect on the behaviour of a metal.

An important type of imperfection is a *dislocation* (fig. 7.4). As its name implies, this is a break in the continuity of the lattice. Very often a plane of atoms will come to a stop. At this point the two neighbouring planes move closer until they are at the correct interatomic spacing. It looks almost as if one plane has been squeezed out. The change from three to two planes causes a disruption of the lattice and leaves a small gap; this is called an *edge dislocation*. Dislocations are an essential element in the forming of metals since, as we will see in chapter 9, they allow plastic deformation to take place at achievable stress levels.

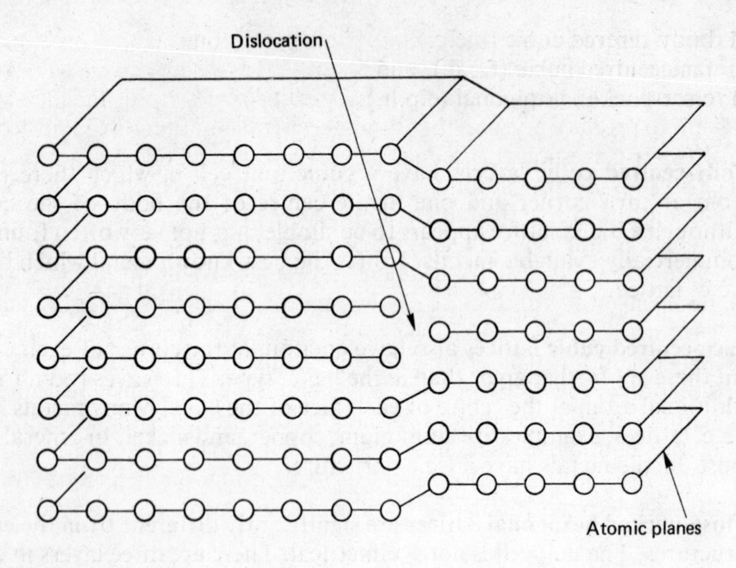

Fig. 7.4 Dislocation in a crystal lattice

Another form of imperfection is a *vacancy* (fig. 7.5). This is literally a hole in the lattice which is there because an atom did not take up the prescribed site. Vacancies play an important role in the diffusion or movement of atoms through the lattice.

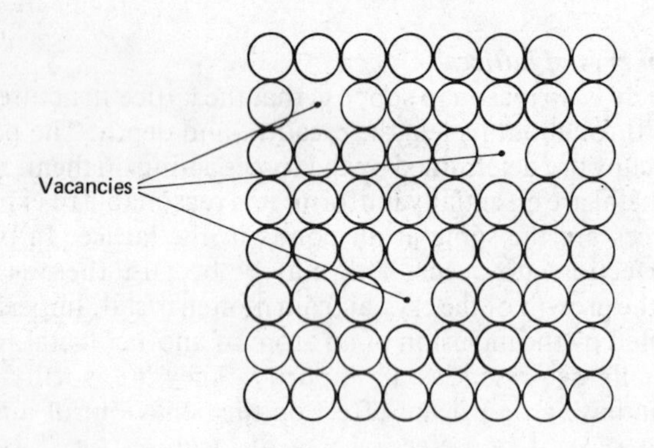

Fig. 7.5 Vacancies in a crystal lattice

7.2 Grain structure

It is possible to produce rods of metal in which the orientation or direction of the lattice planes remains the same throughout the body of the specimen. Each plane goes from one side of the rod to the other without change in direction. Similarly along the length, continuity is maintained from one end to the other. These are identified as *single crystals*. They are very useful in studying the basic properties of a metal, but they are very difficult to produce and are not found in general manufacture. The metals which we use are polycrystalline, i.e. they are made up of a number of small crystals or grains bonded together by interatomic forces. Within each grain, the lattice structure is uniform.

Grains do not have the regular surface appearance we normally associate with crystals. Their shape is determined by interaction with neighbouring grains and can be quite irregular. If we took a slice through a piece of metal – say copper – the boundaries of the grains would appear as a network of lines (fig. 7.6). The grain boundary represents a change in the direction of the lattice planes. As the planes of one grain meet those of a neighbour at an angle, there is considerable mismatch with the result that along the boundary there may be gaps (fig. 7.7). The width of the boundary area is usually equivalent to several atomic diameters, but forces can still act across it to provide cohesion between the grains.

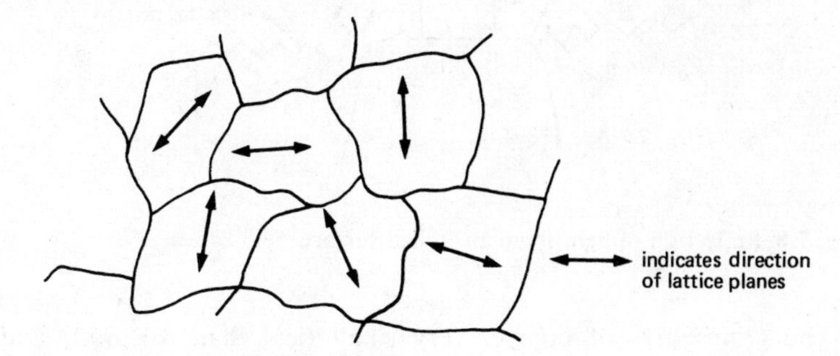

Fig. 7.6 Grain structure of metals

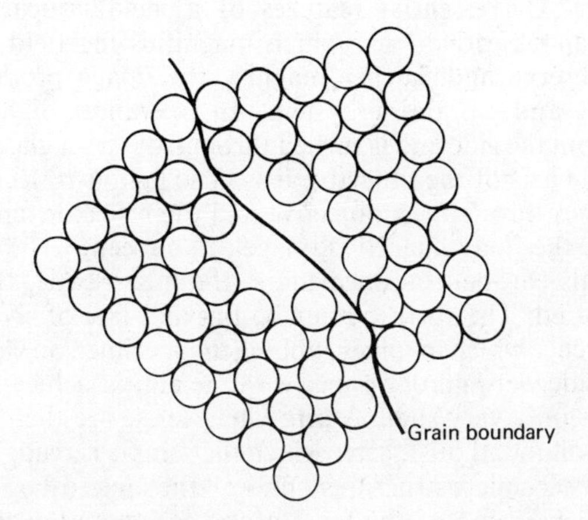

Fig. 7.7 Detail of a grain boundary

Microstructures
These changes in lattice orientation can help us examine the grain structure of a metal. When the surface of a section cut through a piece of metal is attacked with an acid (i.e. it is etched), the atoms at the end of the lattice planes are removed. On a submicroscopic scale, the surface shows ridges corresponding to the edges of the atomic planes. If a beam of light is shone at right angles to the etched surface, it is reflected in different directions (fig. 7.8). Viewed from a position at right angles to the surface, the metal appears to have light and dark areas according to the way the light has been reflected. Grain boundaries which have been attacked by the acid also scatter the light and become visible.

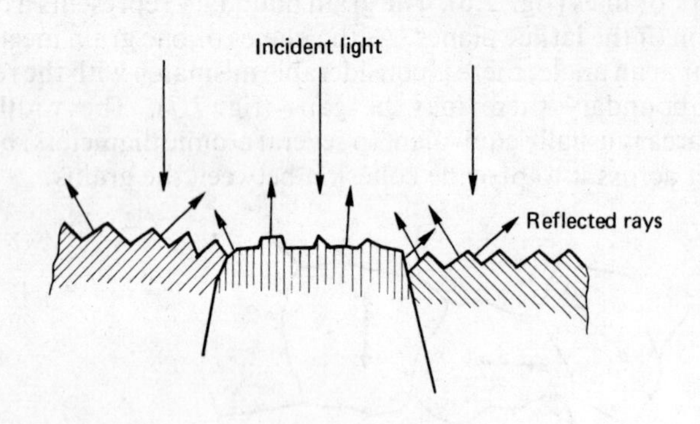

Fig. 7.8 Reflection of light from an etched surface

The grains are, of course, very small (less than 0.1 mm), and a microscope must be used to magnify the view so that the structure can be seen. With a metallurgical microscope, the structure can be magnified up to 1000 times. The essential features of a metallurgical microscope (fig. 7.9) are an objective lens, which magnifies the field of view, an eyepiece to observe and further magnify the image produced by the objective lens, and a lighting system. In operation, light enters the microscope from the side and is turned through 45° by a glass reflector to align it with the axis of the objective lens. The light is reflected from the specimen, passes through the objective and the reflector, and enters the eyepiece where the final magnified image can be seen.

Before metals can be examined under the microscope, they must be carefully prepared. The surface must be flat and free of scratches. This calls for a sequence which involves rubbing the specimen on various grades of silicon-carbide wet-and-dry paper with the aim of achieving a surface which contains only very shallow scratches. These are then removed by polishing with diamond dust, after which the sample is ready for etching.

The study of microstructures under the microscope is called *metallography*. It is one of the most useful techniques available to the metallurgist. Microstructures can give much information about the metal

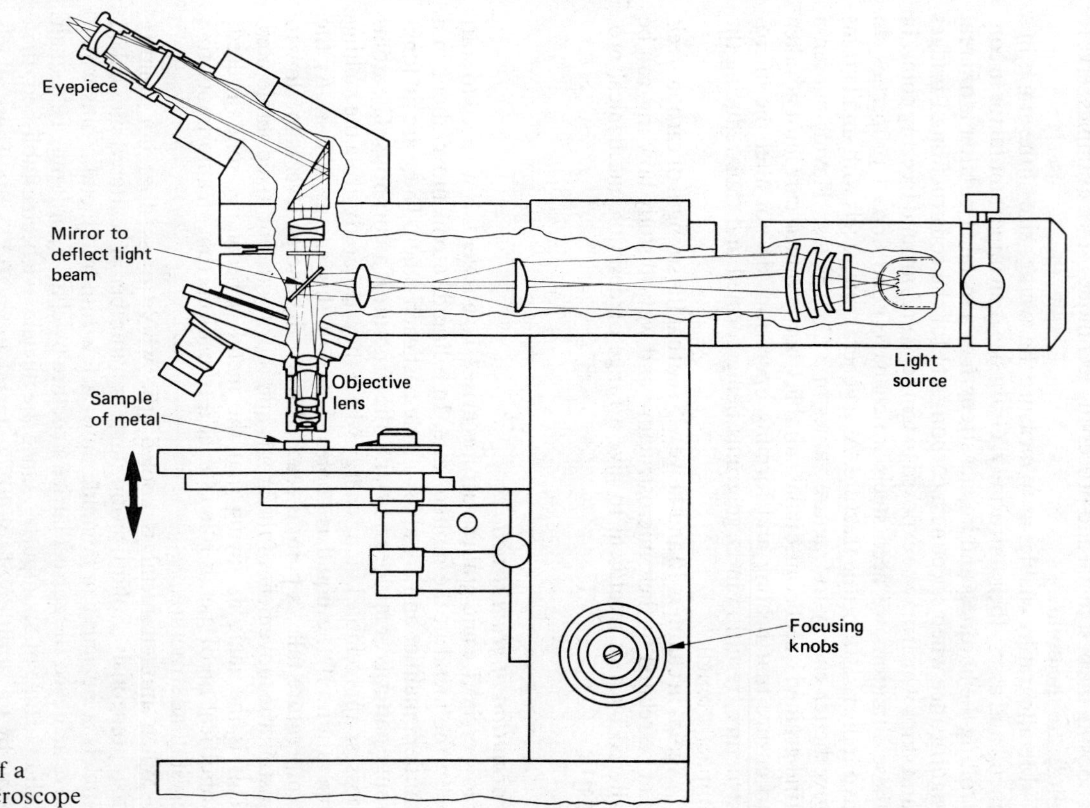

Fig. 7.9 Details of a metallurgical miscroscope

and the treatment it has received. There are important links between microstructure and the properties we can expect from a metal. An understanding of these structure–property relationships is invaluable in choosing the best metal and processing techniques. Two examples serve to illustrate the point.

Metallography enables us to measure the average size of the grains in a metal and assess their uniformity. Grain size is an important factor in controlling the hardness and impact strength of a metal. The latter is of great significance when selecting a carbon steel for the construction of bridges and ships which may be exposed to temperatures around freezing point. In these situations we need steels which have good impact properties, to avoid the risk of brittle fracture. As a general rule, steels with small grains give better energy-to-fracture values in a Charpy test. Having chosen a fine-grained steel because it has suitable Charpy values, care must be taken to ensure that heating and forming operations do not alter the microstructure, resulting in large grains being formed and thus reducing the impact strength.

Again, in chapters 10 and 11 we will see that the strength of carbon steels can be related to their microstructure and we will study how this can be altered by heat treatment to give a range of desirable mechanical properties.

Formation of grain structures

How do lattice and grain structures arise? To answer this we need to start with the metal in the molten state. In a liquid, atoms move about in a random manner and are able to slide past each other. There are no long-range patterns or lattice arrays, i.e. the structure is amorphous. The atoms possess appreciable kinetic energy which they acquired during the melting operation. If the liquid is cooled, some of this energy is lost. As the temperature falls, a point is reached when the energy level is too low to sustain the movement of the atoms and they try to find fixed stable sites. This is the same as saying that the metal is becoming solid. Indeed, solidification of metals is in essence a transition from an amorphous to a crystalline structure.

When molten metal is cooled, the whole mass does not solidify instantaneously. Solidification starts at a number of scattered sites called nuclei. A nucleus can be made up of a few hundred metal atoms which have achieved the correct lattice structure by chance, or it may be a small solid particle of an impurity. Once the nuclei have been established, they grow by the addition of solidifying metal atoms (fig. 7.10). These attach themselves to the nuclei in a regular manner. The first atoms which settle on the nucleus tend to form a column or branch. Subsequently, branches grow at right angles from the main branches to form a *dendrite*. As more atoms attach themselves, the dendrite grows by infilling between the branches and by stretching out further into the remaining molten metal. Growth continues until the outer extremities meet another dendrite. The

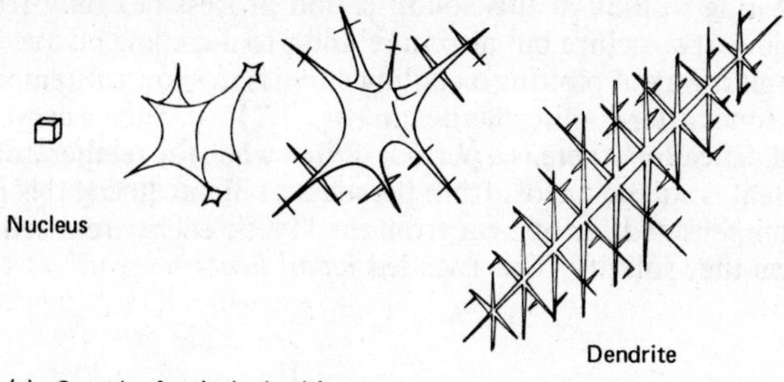

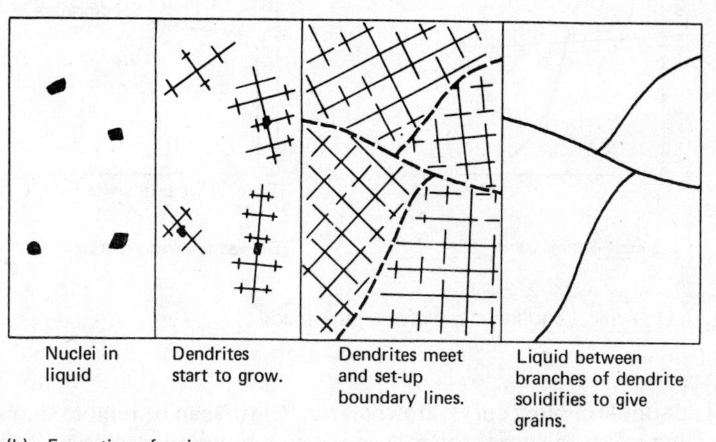

Fig. 7.10 Solidification and the growth of dendrites

branches of this will not be in alignment with those of the first dendrite and a grain boundary is formed.

Although newly solidified atoms always attach themselves in line with or at right angles to the existing planes of atoms, growth after nucleation proceeds to some extent in a random manner and interaction of branches can occur. This is one of the reasons why the metal lattice contains defects such as vacancies and dislocations.

The grain structure, which we discussed in the preceding section, can now be seen to reflect the nucleation-and-growth sequence which takes place on solidification. The size of the grains indicates the number of nuclei which formed as the molten metal cooled. The shape of the grain boundaries is dictated by the rate at which neighbouring dendrites grow.

Cooling curves

It is interesting to look at this solidification process not only from the point of view of structure but also in relation to a cooling curve.

The simplest way of plotting a cooling curve is to show the temperature at various times after cooling has begun (fig. 7.11(a)). Such a curve shows that, at solidification, there is a period of time when the temperature does not fall. Heat is still being lost from the mass of metal during this period, but it is compensated for by heat from the kinetic energy released by the atoms when they solidify. This is called *latent heat*.

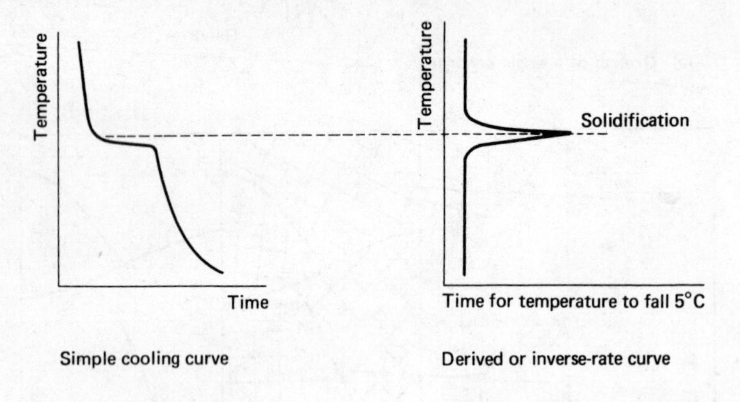

Fig. 7.11 Typical cooling curves for a pure metal

The simple cooling curve shown in fig. 7.11(a) can be replotted to show the time taken to cool through a specified temperature interval, say 5°C (fig. 7.11(b)). The interruption in the cooling curve – i.e. the thermal arrest – which corresponds to solidification shows as a 'spike' in the curve, because at this point it takes much longer to cool through the temperature interval of 5°C. Derived cooling curves plotted in this way give a more accurate indication of the solidification temperature.

Solidification of an alloy

A pure metal solidifies at one temperature which is characteristic of the particular metal. Solidification of an alloy, on the other hand, takes place over a range of temperatures.

An alloy is formed when two or more metals are mixed in the liquid state. In a simple alloy, such as copper–nickel, the atoms of the alloying element (nickel) replace atoms of copper in the lattice. The nickel is said to be soluble in the copper. The amount of nickel added can be increased without changing the situation. If we had started with nickel and added copper, the same structure would have been produced. Copper and nickel are therefore totally or completely soluble in each other in the solid state.

The solidification temperature for a copper–nickel alloy depends on the composition of the alloy. Copper solidifies at a lower temperature than nickel. The effect of adding increasing amounts of nickel to copper is to progressively raise the solidification temperature. The relationship between the nickel content and the temperature at which solidification starts is shown as line ACB on the diagram in fig. 7.12(a). ACB is called the *liquidus*, and at temperatures above this line the alloy is completely molten.

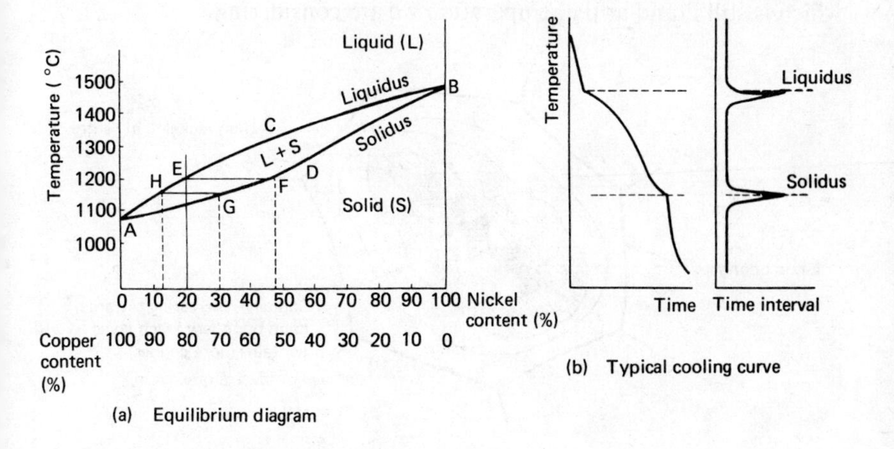

Fig. 7.12 Copper–nickel alloy system

Cooling curves for copper–nickel alloys show two arrests (fig. 7.12(b)). The upper represents the start of solidification (the liquidus) and the lower indicates when solidification is complete (line ADB – the *solidus*). Between these two temperatures there is a mixture of solid and liquid alloy.

The diagram in fig. 7.12(a) thus shows three distinct areas. Above the liquidus an alloy is completely liquid, and below the solidus it is totally solid. Between these two the metal is in a pasty condition as it consists of solid crystals floating in a liquid.

By considering the behaviour of an 80% copper–20% nickel material, we can now see how an alloy solidifies. On the diagram, a vertical line drawn through 20% nickel meets the liquidus at point E. The alloy starts to solidify at this temperature, and the first metal to crystallise out contains both nickel and copper. To discover the composition of this mixture we need to know what alloy would be just completely solid at a temperature corresponding to E. We can do this by drawing a line parallel to the horizontal axis towards the solidus. The intersection of this line and the solidus (F) gives us the composition of the nickel–copper mixture which solidifies first. For our 20% nickel material, it would be 48% nickel–52% copper.

As the temperature falls, the next metal to solidify contains more copper and again we can predict its composition by drawing a horizontal line to meet the solidus, e.g. at point G. Note that the copper content is still less than 80%. This means that the molten metal adjacent to the dendrite must be becoming richer in copper. If we want to know how much copper is in the remaining liquid, we can extend the line in the opposite direction until it intersects the liquidus (at H). This gives us the composition of the alloy which is still liquid at the temperature we are considering.

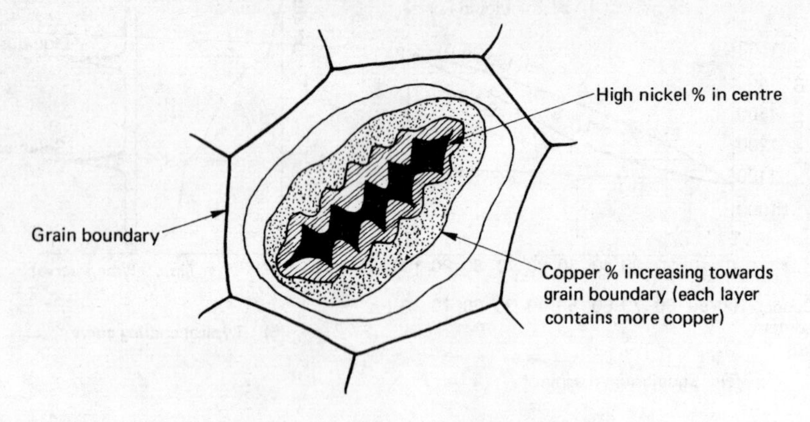

Fig. 7.13 Coring in a 80% copper–20% nickel alloy

As a result of this sequence, the dendrite contains layers of different composition (fig. 7.13). The centre is high in nickel and low in copper, whereas the outer layer is high in copper and contains more than 80% of this element. There is thus a composition gradient across the grain, but it is important to recognise that the *average* composition remains at 80% copper–20% nickel. Such *cored structures* are often found in metals where the cooling rate has been reasonably fast. If, on the other hand, the solidification rate is slow, copper atoms move through the lattice from the outer layers towards the centre of the dendrite to give a grain of uniform composition. When this happens, the grain has reached a stable or *equilibrium* condition. Although we use the diagram in fig. 7.12 to illustrate the formation of cored structures, it is in reality an equilibrium diagram and is strictly accurate only when the cooling rate is sufficiently slow to allow movement or diffusion of the copper atoms to occur.

7.3 Equilibrium and change

These related concepts of equilibrium and diffusion are very relevant to our understanding of how metals behave. Metallurgical changes or reactions, as they are often called, need time to take place. Many of these changes rely on the mechanism of diffusion, which is simply a movement of atoms through a lattice structure. In the cored grain described above, a composition gradient existed as a direct result of cooling too fast for equilibrium conditions to be set-up. Wherever there is a gradient of this type, atoms always try to move from an area of greater concentration into one where there is less alloying element present. As long as this difference in concentration exists, the grain is in an unstable condition and only reaches equilibrium when the alloying atoms have been uniformly distributed. At room temperatures, there is insufficient energy available for the atoms to move through the lattice. On the other hand, at temperatures approaching the melting point, the atoms can move rapidly. The rate of diffusion is therefore temperature-dependent.

The differences in properties produced under equilibrium and non-equilibrium conditions can be quite dramatic. For example, a steel containing 0.4% carbon has a hardness of 190 HV when cooled slowly, but this increases to over 500 HV when a hot sample is plunged into cold water. When interpreting metallurgical data, we must always take care to identify the conditions under which they were determined.

7.4 Equilibrium diagrams

An equilibrium diagram records the changes which take place in an alloy when the cooling rate is sufficiently slow to allow metallurgical reactions to go to completion. It does this for all the possible combinations of two or more elements in a system. A binary diagram describes a system containing only two elements, e.g. the copper–nickel alloys. If there are three elements in the system, a ternary diagram is used. Here we will be concerned only with binary systems.

The equilibrium diagram is, in effect, a plan or a map which enables us to find out what the microstructure is at any particular temperature. The diagram shows the *phases* which are present – these are areas or regions which differ from one another and can be distinguished under the microscope. Phases may have different compositions or they may show a change in structure, say from a solid solution to a compound in which the atoms of two or more metals are bonded together in a cluster rather than in an orderly lattice.

Although some equilibrium diagrams are relatively simple, as for example the copper–nickel diagram, many are very complex and are difficult to interpret. Fortunately, we do not need to understand the whole of a diagram to make use of it: we can use it selectively. For example, it is possible to predict the phases which will be present at room temperature without necessarily knowing how they got there or what happened during the cooling from the solidus. In many instances we can confine our study solely to one composition and derive information about the temperatures to be used for heat-treatment operations.

In subsequent chapters, we will be using equilibrium diagrams to help us understand the behaviour of various commercially used alloys. The diagrams will often contain two significant features: a eutectic and a partial or limited solubility of one metal in another. We must know what these are before we can successfully interpret equilibrium diagrams.

Eutectics

Earlier in this chapter it was noted that, in general, the solidification of an alloy takes place over a range of temperatures. There is an exception to this. In many alloy systems there is a particular composition – the *eutectic composition* – at which solidification occurs at a single temperature – the *eutectic temperature*. There is then no solidification range for this composition.

The eutectic composition can be readily identified on an equilibrium diagram because it has the lowest melting point of any alloy in the system. The liquidus shows a trough with the eutectic at the bottom. We can also note that the solidus temperature is the same for all alloys in the system and corresponds to the eutectic temperature.

The formation of a eutectic can be seen in the lead–antimony alloys (fig. 7.14) which are used to line crankshaft bearings. The eutectic contains 13% antimony and solidifies at 240°C. In alloys which have less than this amount of antimony (e.g. alloy X), the first metal to solidify is lead. This forms dendrites which continue to grow as the temperature falls. Lead is therefore taken from the liquid, which changes in composition until it contains only 87% of this element, i.e. the antimony

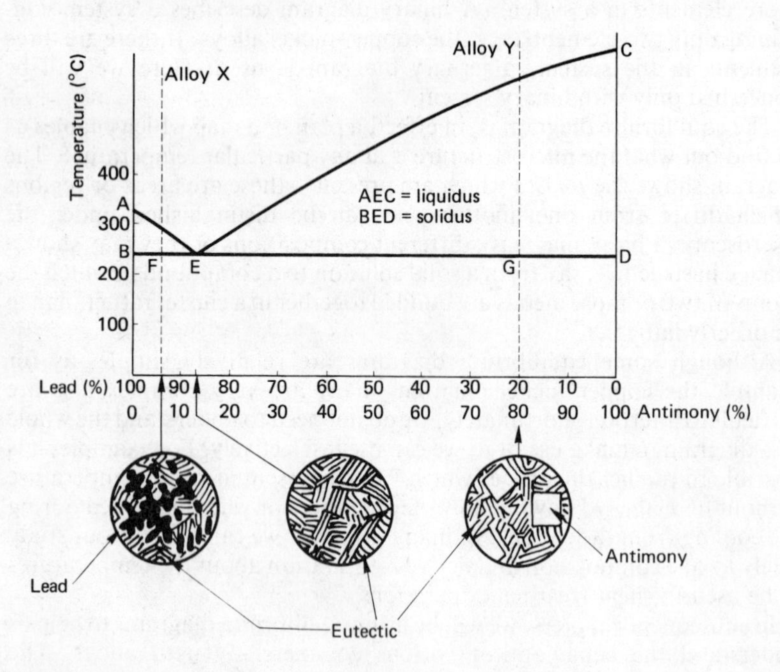

Fig. 7.14 Lead–antimony system

content has risen to 13%. This is the eutectic composition and the temperature has fallen to the solidus, below which the metal cannot be molten. At the solidus, the lead and antimony in the remaining liquid solidify simultaneously and form a laminated or layered structure. The solid metal contains lead dendrites surrounded by lead–antimony eutectic.

If the alloy has less than 87% lead (e.g. alloy Y), the first metal to solidify is antimony and the structure is made up of cubes of this metal embedded in the eutectic.

The relative amounts of pure metal (lead or antimony) and the eutectic in the solidified metal are given by the intersection of the vertical composition line and the solidus:

$$\text{for alloy X,} \quad \frac{\text{mass of lead}}{\text{mass of eutectic}} = \frac{FE}{BF}$$

$$\text{for alloy Y,} \quad \frac{\text{mass of antimony}}{\text{mass of eutectic}} = \frac{GE}{DG}$$

Antimony is hard compared with lead, and the combination of wear-resistant particles in a soft matrix is an ideal structure for a bearing metal. A typical bearing alloy contains 20% antimony.

Partial solubility

So far we have studied two alloy systems: copper–nickel and lead–antimony. The first obvious difference between the two is the absence of a eutectic in the copper–nickel system. Another difference is perhaps not so readily apparent – at room temperatures, lead and antimony do not dissolve in each other to any significant extent, whereas copper and nickel do.

In a solid solution, the solute atoms replace some of the solvent in the lattice structure. (When we are talking about solutions, the *solute* is the metal which is taken into solution, while the *solvent* is the metal in which the solute is dissolved.) The number of solute atoms which can be accommodated in the lattice depends partly on their size in relation to the solvent atoms. With a few systems, the atoms of the two metals are completely interchangeable in the lattice. This applies to the copper–nickel alloys, in which the metals are completely soluble in each other. More usually, the difference in the size of the atoms leads to a distortion of the lattice. Only a certain amount of distortion can take place, and when this point is reached no more solute atoms can be accommodated – i.e. the solubility is limited.

The extent of this *partial* solubility depends on the alloy system but is also influenced by the temperature. As a metal is heated, it expands and the atoms move further apart. This means that there is more space to accommodate size differences in solute atoms and the solubility increases.

The lead–tin alloys (fig. 7.15), which are principally used as solders, exhibit partial solubility. The solid solutions are indicated on the

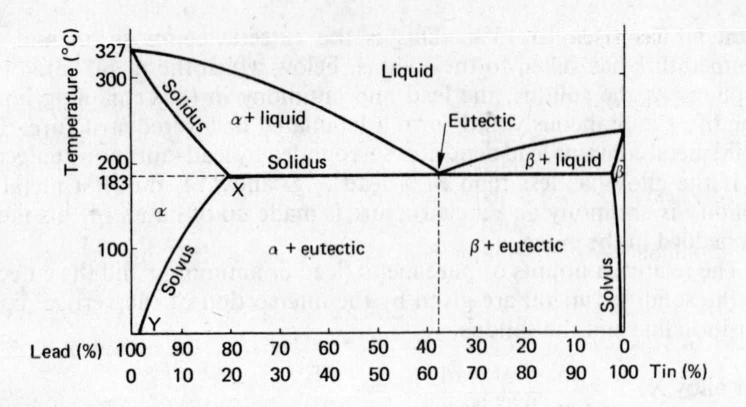

Fig. 7.15 Lead–tin system

equilibrium diagram by the use of Greek letters. The solution of tin in lead is designated α (*alpha*), while that of lead in tin is β (*beta*).

The amount of tin which can be dissolved in lead at temperatures below the solidus is shown by the line XY, which is called the *solvus*. At 183°C, up to 20% of tin can be accommodated in the lead. This falls to 3% at room temperatures. The tin which is rejected from the lead lattice sets up its own lattice into which some lead is dissolved. That is, it forms β which is added to any of this phase already existing in the alloy.

The structure of alloys with compositions between the solubility limits, i.e. between 97% lead–3% tin and 1% lead–99% tin, is a mixture of α and β phases. In alloys with less than 62% tin (the eutectic composition) the structure is α plus eutectic, whereas above this tin content it is β plus eutectic. The eutectic itself is made up of layers of α and β phases which have solidified simultaneously from the melt at 183°C.

The information provided by the equilibrium diagram about the solidification range (i.e. the temperature difference between liquidus and solidus) is useful in choosing the correct solder. Plumber's solder contains 65% lead, and the diagram shows that it has an appreciable solidification range during which the metal consists of solid α phase in a liquid. During solidification, the metal is pasty and is well suited to the soldering of joints in pipework since it allows the plumber ample time to shape the joint by wiping it with a cloth or chamois leather. Solders for electrical work, on the other hand, must solidify as quickly as possible without having to cool through a weak pasty range. The solidification range must therefore be as short as possible, and the equilibrium diagram shows that the eutectic is the best choice, especially as it has the added advantage of a low soldering temperature which minimises the risk of heat damage to components.

Some of the terms used in connection with equilibrium diagrams are summarised in Table 7.1.

Table 7.1 Terms used in equilibrium diagrams

Term	Explanation
Alloy	A substance containing two or more metallic elements which may be simply mixed, or in solution, or combined.
Alloy system	A series of alloys representing a range of compositions based on, say, two specified metallic elements.
Solid solution	A uniform mixture of two solid metals. In a solid solution, the atoms of the solvent metal are replaced in the lattice by those of the metal which is dissolved (i.e. the solute).
Phase	A uniform portion of an alloy system. The molten metal in the system represents one phase. In the solid state, there may be either one or two phases present. Solid solutions are examples of phases. A solid solution in which one of the pure metals is the solvent is called a *primary phase*.
Equilibrium	A state of balance in an alloy when the structure has achieved a stable condition.
Equilibrium diagram	A diagram showing the phases present in an alloy at temperatures from 0°C to the highest melting point when all structural changes have been allowed to go to completion. Equilibrium diagrams are constructed from data collected in a large number of experiments designed to detect phase changes.
Liquidus	The line on an equilibrium diagram indicating the temperatures at which solidification starts during the cooling of a series of alloys.
Solidus	The line defining the highest temperatures at which the alloys are completely solid.
Solvus	The line showing the maximum solubility of one metal in another at different temperatures.
Eutectic temperature	The lowest melting point in an alloy system. The eutectic temperature is lower than the melting points of either of the pure metals.
Eutectic composition	The alloy composition which gives the lowest melting temperature.

8 Metals for casting

One of the distinguishing features of a liquid is its ability to take up the shape of the vessel containing it. Molten metal poured into a container or mould flows into the corners and fills all the voids. If the metal is allowed to solidify, the shape of the resulting solid is similar to that of the mould. Note that it is *similar* but not *exactly the same*, because solid metal is denser than molten metal and at solidification there is a reduction of volume which varies from 2% to 9% for commercial alloys. When this casting technique is used to make metal parts, the shape of the cavity in the mould is adjusted to allow for the shrinkage, so that the product has the correct dimensions.

Hollow components can be produced by inserting a *core* into the cavity in the mould. The molten metal flows around the core, which is removed from the casting after the latter has cooled.

8.1 Why use casting?

Casting offers a number of advantages and is widely used in the manufacture of components ranging in mass from a few grams up to several tonnes. One of its most attractive features is the ability to form a shape in one operation. This not only represents a considerable saving in labour costs but also avoids any problems which might arise in joining sub-assemblies together. As an example, some of the alloys used in chemical plant can be difficult to weld. In these cases, it is useful to be able to cast the pump bodies and pipe fittings and to incorporate flanges which can be bolted to make connections to other components. It is also often possible to produce accurate castings in metals which are difficult to machine to size. In this way, the machining operation can be completely eliminated or, at worst, only a light grinding is needed to achieve the final dimensions.

A further important advantage of casting is that a high degree of reproducibility is possible. Once the shape of the mould has been established and the alloy composition has been fixed, control of the melting and pouring conditions ensures that all castings are identical. This has many obvious attractions in the volume production of a range of items.

Even with simple shapes in easily fabricated metals, casting can offer cost savings. To a large extent, this is limited by the number of components to be produced. The cost of preparing a mould can be very high, especially for hollow components which require cores. This cost must be recovered against the job on hand, and the mould cost per component is reduced as more of the same units are cast. For any given design there is a point at which casting becomes cheaper than fabrication. With large press frames it is sometimes worthwhile casting even in the one-off situation. On the other hand, with intricate machine components a production run of several hundred may be needed to offset the initial costs. Such analyses are seldom straightforward and are critically dependent on the method of casting which is used.

8.2 Principal methods of casting

There are four principal methods of casting metals which we need to consider:

a) sand casting,
b) investment casting,
c) gravity die-casting,
d) pressure die-casting.

The first and second of these use a mould made from refractory (i.e. heat-resistant) non-metallic materials, while the mould for die-casting is machined from metal blocks.

Sand casting (fig. 8.1)
With this method of casting, the mould is made by ramming sand around a pattern which has the same shape as the component but includes an allowance for shrinkage. When the mould is removed, it retains an imprint of the component's shape. The mould is made in two halves which are placed together.

The patterns are built up on a flat board or plate. Frequently they are constructed from wood, but they can be machined from aluminium or possibly steel. In some cases, an expanded-polystyrene pattern can be used. This is left in place in the mould, where it is decomposed and replaced by the molten metal.

A moulding box is placed around the pattern to retain the sand and to make it easier to handle the mould. The sand needs to be bonded so that it keeps the shape of the component when the box is lifted off the pattern. Various bonding agents can be used; the most common are clay, sodium silicate, and cold-setting resins.

The top half of the mould must contain a hole for the metal to be poured in (i.e. a feeder). Further holes – called *risers* – are needed to provide a reservoir of liquid. This is used to feed molten metal to the casting during the solidification process, thus compensating in part for the volume shrinkage which takes place.

Before the mould is put together, the cores, if needed, are inserted. These must be strong enough to support themselves. For this reason they are usually made from a resin-bonded sand.

The two halves of the mould are located accurately and the boxes are clamped together. Metal is poured into the feeder(s) until it flows up the risers, showing that the mould is full. When the casting has cooled to a temperature well below the solidus, the sand mould is removed by placing

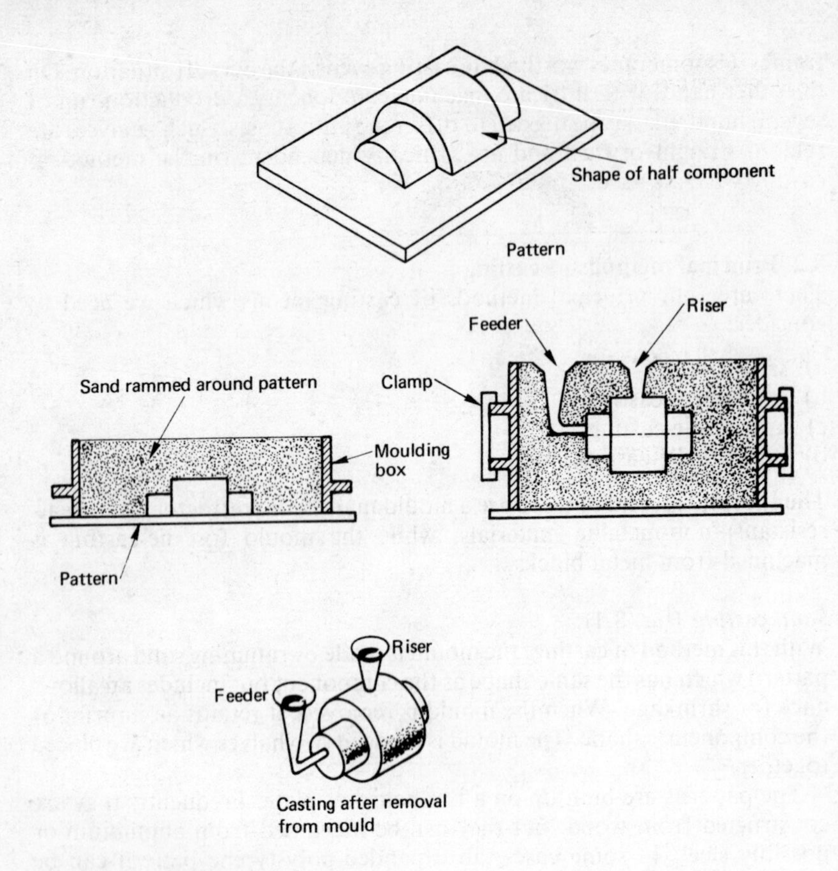

Fig. 8.1 Sand casting

it on a vibrator. The sand is shaken off the casting and, after sieving to remove fused particles, is used for further moulds.

Sand casting is versatile and is suitable for a variety of complex shapes. Moulding costs are not high in comparison with those for the metal moulds used in die-casting. The production costs, even for a small number of units, make the process competitive with other methods of fabrication. Against this, the surface finish is generally poor, although some improvement can be achieved by shell moulding (fig. 8.2). In this variant of sand casting, the mould is made in sand which is bonded with a thermosetting resin. This gives a smoother surface to the cavity and, at the same time, imparts greater strength to the sand – which means that the mould can be in the form of a thin shell.

Another drawback to sand casting is that up to 35% of the metal poured into the mould ends up as feeders and risers which are scrapped. This scrap metal must be capable of being recycled if the best economy is to be achieved.

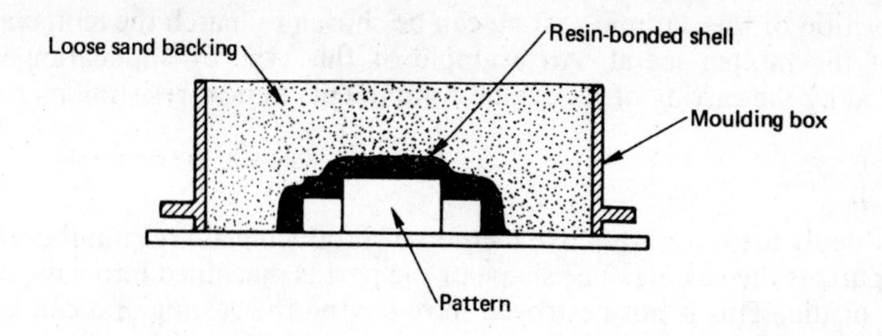

Fig. 8.2 Shell moulding

Investment casting

Small parts, such as machinery components (fig. 8.3), can be accurately produced by investment casting in a mould made from a ceramic.

A replica of the part is shaped in wax – making allowance for the predicted shrinkage during solidification and cooling, so that the finished casting has accurate dimensions. A slurry of ceramic, such as fire-clay, is coated on to the surface of the wax. When the coating has set, the wax is melted and poured away leaving a cavity. Usually a number of these moulds are linked together and attached to a common feeder.

Investment casting is slightly more expensive than sand casting for small-batch production, but it gives a superior finish. It is particularly suited to the casting of metals which have a high melting point, since the

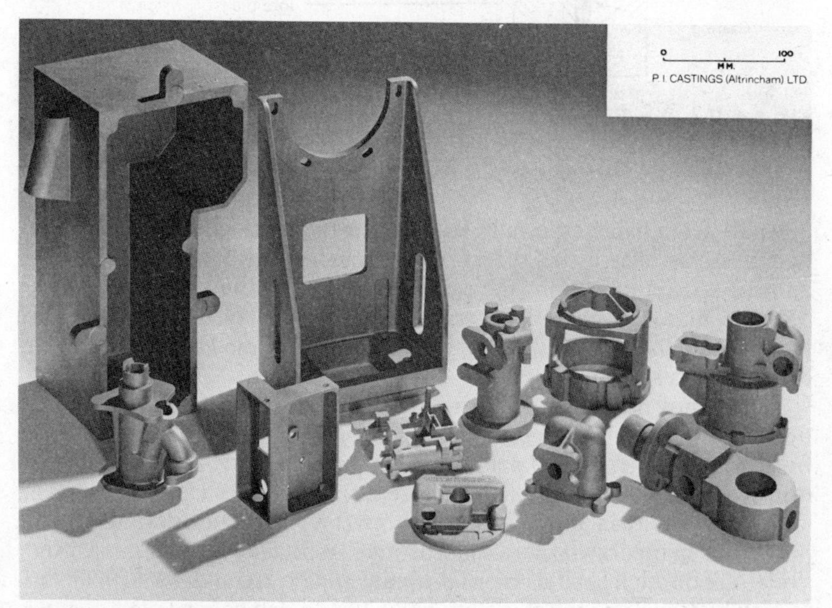

Fig. 8.3 Investment castings

composition of the ceramic coating can be chosen to match the temperature of the molten metal. An example of this type of application is provided by the casting of jet-engine components in heat-resisting nickel alloys.

Die-casting

Undoubtedly the most attractive method of producing a large number of metal parts is die-casting. The shape of the part is machined into a metal mould or die. This is not destroyed in removing the casting and can be repeatedly reused until its dimensions have been changed by wear and erosion.

There are two types of die-casting. In *gravity die-casting* the metal is poured into the mould in much the same way as it is in sand casting. The liquid freezes quickly in the metal mould and only relatively simple shapes can be produced. With *pressure die-casting* (fig. 8.4), on the other hand, the liquid metal is forced into the die, which may be heated to ensure that it is properly filled before solidification begins.

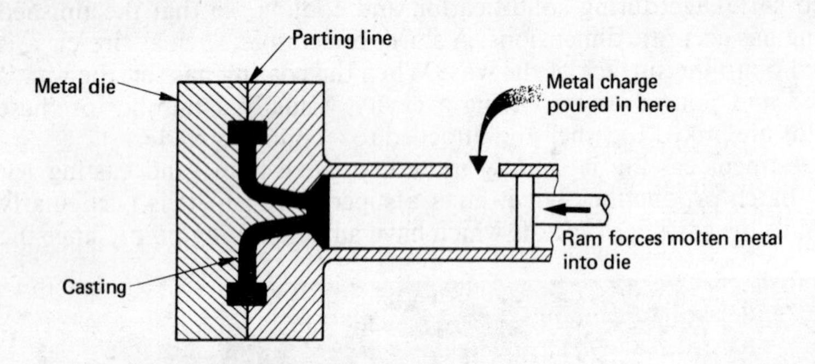

Fig. 8.4 Pressure die-casting

Apart from being especially suitable for the production of large numbers of items, die-casting offers the great benefit of good surface finish, and for most applications the castings do not require machining. A car contains many die-cast units - carburettor bodies, inlet manifolds, fuel pumps, and door handles are but a few examples of die-castings. A typical die-casting is shown in fig 8.5.

The method has an added advantage in that metallic inserts can be positioned in the mould before pouring and can thus be incorporated into the finished component. Steel pins can be set into a gearbox cover to aid location when it is being installed. Bronze bearing bushes can also be positioned accurately in an end plate to support a rotating or sliding shaft.

The two main drawbacks to die-casting - especially the pressurised version - are the high capital cost of the plant and the die and the difficulty of casting high-melting-point metals which would damage the die.

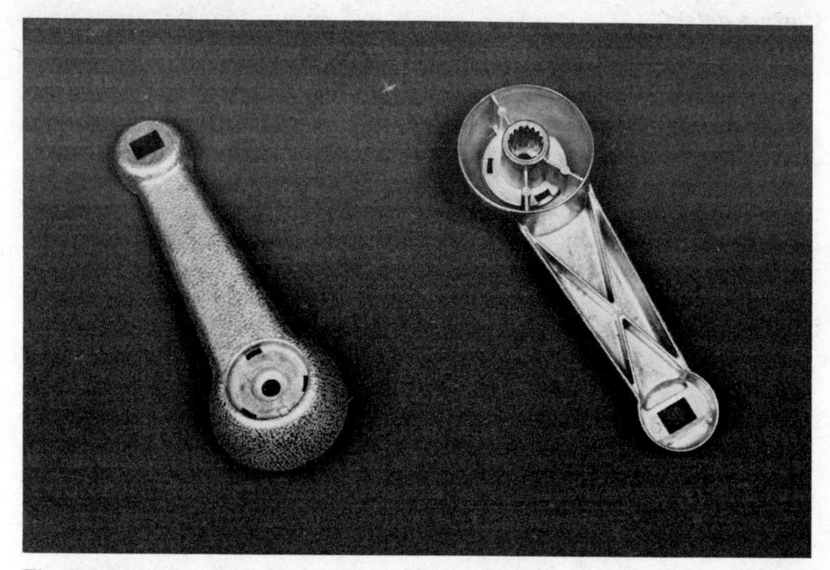

Fig. 8.5 Window-winder handle for the Ford Fiesta car, pressure die-cast in zinc alloy

8.3 Properties of castings

In general, castings have poorer mechanical properties than components produced by other methods of forming. This is not to say that they are unsatisfactory in service: on the contrary, a well chosen combination of alloy and casting technique will give acceptable properties.

The properties of a material when it is cast are influenced by two important factors:

a) grain structure,
b) the presence of defects.

Grain structure

With most casting techniques, conditions are far from equilibrium. Cooling rates in die-casting can be very fast; the slowest cooling is found with sand moulds. We saw in the previous chapter that coring occurs when solidification takes place under non-equilibrium conditions. The extent of this coring and its effect on mechanical properties depends on the particular alloy.

The cooling rate also affects the size of the grains. The slow cooling of sand casting gives large grains which reduce the tensile and impact properties. Rapid solidification during die-casting, on the other hand, leads to nucleation at a large number of sites and results in finer grain size (see page 100). Consequently, die-castings tend to have better strength than sand castings in the same alloy.

Casting defects

There is always a risk that defects will be present in the finished castings if the operation has not been properly controlled. These defects arise from a variety of sources, such as moisture in the sand, loose material on the surface of the mould, incorrect pouring techniques, and inadequate filling of the risers. The most common defects are as follows.

a) *Porosity* – holes in the casting, which can reduce its pressure tightness.
b) *Shrinkage* – a term used to describe areas which appear to have a spongy structure and are often located just below the risers. Shrinkage reduces pressure tightness and strength if it is extensive.
c) *Inclusions* – sand, oxides, or non-metallic slag from the melting operation trapped in the body of the casting. Inclusions can cause problems in machining.
d) *Cracks* always reduce strength, but they may be removed by machining if they are confined to the surface.
e) *Surface faults* are normally removed by machining, but they can be a problem in die-castings where it is intended that the components should be used in the as-cast condition.

Importance of alloy selection

The defects mentioned in the previous section should not be present if the casting process is carried out correctly. The ease of control is influenced, however, by the choice of alloy and the design of the component. Apart from the desirability of using a low-melting-point material wherever this is possible, the alloy should have good fluidity (low viscosity) so that it can flow into the corners of the mould cavity and fill it completely. The metal should also have good strength at the solidification temperature, to minimise the risk of hot cracking or tearing. This in turn is partly dependent on the design. Cracking results from the stresses set up by shrinkage during solidification and subsequent cooling. The design of the casting should avoid details which cause stress to be built up in local areas.

Various details which cause problems are illustrated in fig. 8.6, where some other aspects of design of castings are also summarised.

8.4 Typical casting alloys

It would be difficult in the short space of this chapter to give a complete list and description of all the metals which are suitable for casting. The following selection has been made to illustrate some of the more important characteristics of different alloy systems.

Aluminium–silicon alloys

Aluminium has high electrical conductivity and good corrosion resistance. With a relative density of 2.7, the mass of a piece of aluminium is only one third that of the same volume of steel. This is often the main reason why aluminium components are chosen for aircraft and road vehicles, where the reduction of weight is a major consideration.

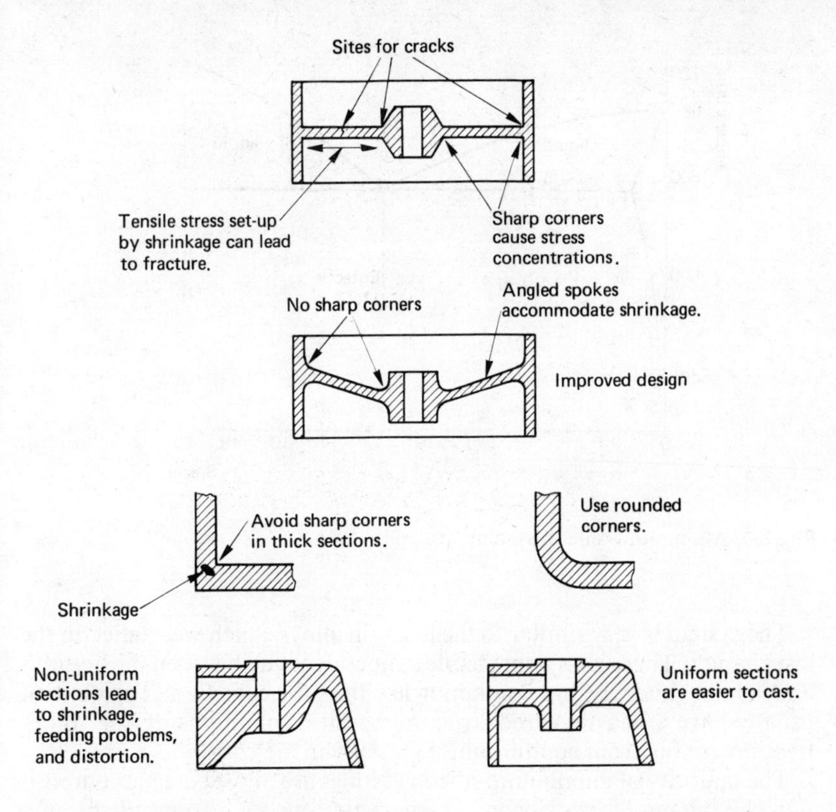

Fig. 8.6 Some aspects of design of castings

The pure metal has low strength in the as-cast condition, i.e. when it has a dendritic structure. For this reason, it is usually alloyed with other metals. Some specialised high-strength alloys have been devised especially for use in aircraft, but most commercial castings are made from an alloy of aluminium and silicon.

As soon as aluminium is exposed to air, an oxide skin is formed on the surface of the metal. This gives the solid its excellent resistance to corrosion, but it also imparts high surface tension to the liquid metal. In both sand casting and die-casting, this poses problems as the liquid tends to pull away from the corners of the mould. The addition of 10% to 13% of silicon to aluminium produces an alloy which has lower surface tension and better fluidity than the pure metal. This alloy is equally well suited to sand casting, gravity die-casting, and pressure die-casting.

The equilibrium diagram shows that silicon has only a limited solubility in aluminium which falls from 1.65% at the solidus to 0.05% at 20°C. There is also a eutectic at 11.7% silicon. Alloys containing more than 15% silicon are of little commercial use, and the diagram in fig. 8.7 goes only to 20%.

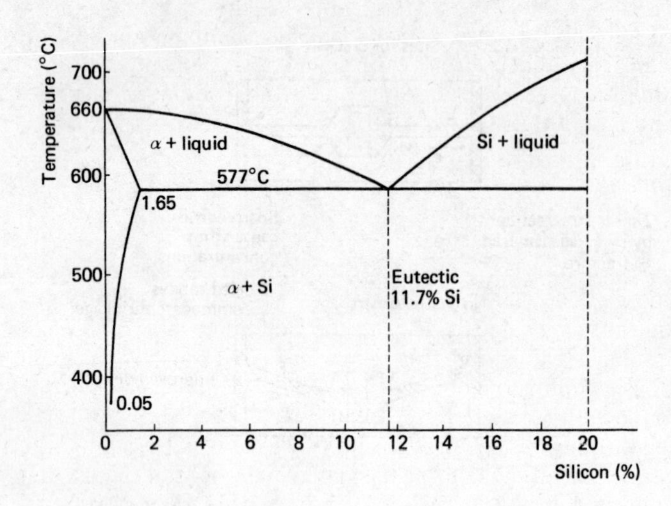

Fig. 8.7 Aluminium–silicon system

The system is very similar to the lead–tin alloys which we studied in the last chapter. There is an appreciable temperature gap between the liquidus and solidus, and castings containing less than the eutectic silicon content usually have a heavily cored structure because conditions during solidification are far from equilibrium, especially in die-casting.

The majority of aluminium–silicon castings are of the eutectic composition, i.e. about 11.7% silicon. The eutectic consists of pure silicon in a matrix of aluminium which has only a small amount of silicon dissolved in it (α phase). With a simple aluminium–12% silicon alloy the eutectic has a coarse structure, resulting in low strength and ductility. Some improvement can be achieved by faster cooling, but this may not be easy to arrange. A better method of refining the eutectic structure is to modify the alloy by adding a very small amount of sodium to the molten metal just before it is poured into the mould. Only about 0.01% sodium is needed to bring about a change in the microstructure. The grain size is drastically reduced, and both the strength and ductility are improved (Table 8.1).

Table 8.1 Effect of modifying an aluminium–silicon alloy (10% to 13% silicon; BS 1490:1970 alloy type LM6)

Condition	Tensile strength (N/mm^2)	Elongation (%)
As cast	125–140	5
Modified	185–200	7

As a result of their good fluidity, aluminium–silicon alloys can be used to pressure die-cast thin sections. This is particularly useful in the manufacture of automobile crank cases, where the combination of thin walls and ribs can give a light but rigid component. Other examples of the use of aluminium–silicon castings are inlet manifolds, pump bodies, gearbox covers, and mounting or pedestal brackets.

Magnesium alloys

Another material which has a low density is magnesium. Its relative density is only 1.74 and it is the lightest casting material in commercial use. Pure magnesium has very low strength and must be alloyed for most practical purposes. The most widely used casting alloys contain zinc and zirconium. The latter acts as a grain refiner by encouraging more and therefore smaller grains to be formed, thus giving improved ductility.

The alloys exhibit very good machineability. Much is made of the fire-risk when machining magnesium, but given the observance of a few well established precautions this presents no problem in modern manufacture.

In general, the mechanical properties of magnesium die-castings are good, but impact values can be poor in comparison with aluminium. The low Young's modulus (44×10^3 N/mm^2) means that stiffeners must often be designed into the casting to limit deflections in thin sections.

Copper–tin alloys

Bronze is probably one of the oldest commercially produced metals. It is an alloy of copper and tin – castings usually contain 7% to 10% of the latter element.

The equilibrium diagram (fig. 8.8) is complex and shows that a number of phase changes or transformations occur during cooling from the

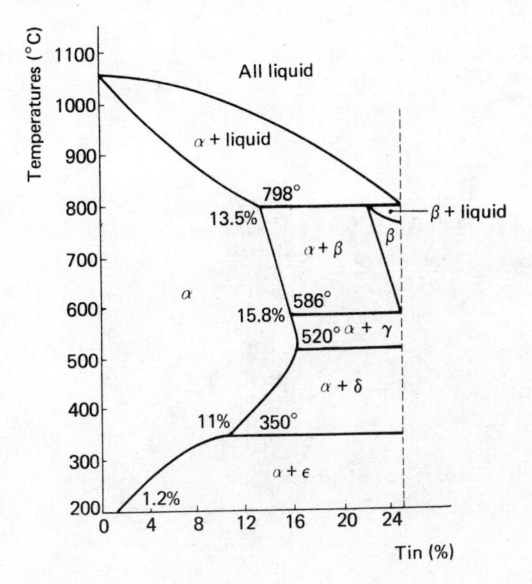

Fig. 8.8 Part of the copper–tin system covering bronzes

solidus. For our present purposes we can ignore the changes which take place at higher temperatures and consider the δ (*delta*) and ϵ (*epsilon*) phases. The rate of transformation from δ to ϵ is very slow and in castings it rarely has time to take place. The structure at 350°C is therefore retained at room temperature, and a bronze containing 10% or less of tin should be entirely composed of α phase. This is a solid solution of tin in copper and has good ductility.

It will be noted, however, that the temperature interval between the liquidus and solidus is 170°C for a 10% tin alloy. This leads to a significant amount of coring, and the last metal to solidify is rich enough in tin to contain small amounts of δ phase. This phase consists of an intermetallic compound which has the formula $Cu_{31}Sn_8$. Metallic compounds are quite different from the solid solutions we have discussed so far. Solid solutions retain the characteristic structure of the solvent metal and accommodate varying amounts of the alloying atoms which occupy sites within the lattice. Metallic compounds, on the other hand, consist of clusters of atoms in which the two metals are present in a fixed ratio. The δ phase in bronze contains atoms of copper and tin in the ratio 31:8.

Metallic compounds are usually hard and brittle and are present as particles which can be seen under the microscope. In the tin bronzes, the δ phase is embedded in the α solid solution (fig. 8.9). This is the ideal combination for a bearing material in a motor or a piece of machinery. The ductile α phase withstands any shock loading, while the hard δ phase resists wear.

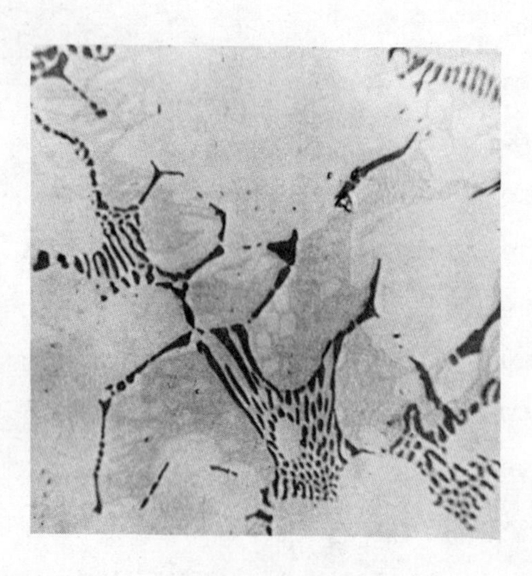

Fig. 8.9 Microstructure of tin bronze (magnification $\times 100$)

Bronze castings are rarely a simple copper–tin alloy: other elements are added to achieve specific properties. Three of the most important are zinc, lead, and phosphorus.

a) *Zinc* improves the casting characteristics and reduces the cost of the alloy by replacing some of the tin, which is more expensive. An 88% copper–10% tin–2% zinc alloy used to be known as Admiralty gunmetal. It is no longer used for gun barrels but gives castings which have good corrosion resistance coupled with strengths of about 300 N/mm^2.
b) *Lead* confers improved machineability on the casting. Usually 2% is sufficient for this purpose. Larger quantities (up to 25%) can be added to bearing bronzes to assist in lubricating the surfaces where access for maintenance is a problem.
c) *Phosphorus* in the first instance removes oxygen from the molten metal and in doing so helps to eliminate porosity. Phosphor-bronze bearings also have improved wear resistance.

Although bronze is a relatively expensive material, it is often selected for applications other than bearings because it offers a combination of high strength, reasonable thermal conductivity, and good corrosion resistance. Pumps and valves in petrochemical plant and in hydraulic systems provide examples of bronzes being used to good effect.

Zinc die-casting alloys

For most people, zinc means galvanising. Steel sheets coated with a thin layer of zinc have a high resistance to atmospheric corrosion, especially in moist conditions. However, about 20% of the zinc produced commercially is used for die-casting small components, especially for the car industry. Frequently these are chromium-plated and are not readily recognisable as a zinc alloy. On the other hand, the bodies of carburettors are used in the as-cast condition and illustrate the high degree of surface finish and accuracy which is a characteristic of die-casting.

The zinc used for the alloys must be of very high purity. In particular the iron, lead, cadmium, and tin contents must be kept to very low levels. If this is not done, the casting suffers from corrosion which penetrates around the grain boundaries, i.e. the attack is intercrystalline or intergranular. The material then looses its ductility and fractures easily.

Two alloys are in current use: British Standard BS 1004 types A and B (Table 8.2). Both contain 3.9% to 4.3% aluminium and 0.03% to 0.06% magnesium. Type B also has 0.75% to 1.25% copper. The alloys have a low melting point, low surface tension, and good fluidity. They are thus ideally suited to die-casting.

Type A is the most commonly used alloy and has a tensile strength of 280 N/mm^2. Type B has a higher strength (330 N/mm^2) as a result of the copper addition, but is significantly less ductile.

Zinc die-castings shrink over a period of time. Alloy A loses 3.0×10^{-4} mm in each millimetre during the first five weeks after casting, rising

Table 8.2 Zinc die-casting alloys (BS 1004:1972)

	Alloy designation	
	BS 1004A	BS 1004B
Composition %		
aluminium	3.9–4.3	3.9–4.3
copper	0.03 max.	0.75–1.25
magnesium	0.03–0.06	0.03–0.06
iron	0.075 max.	0.075 max.
lead	0.003 max.	0.003 max.
cadmium	0.003 max.	0.003 max.
tin	0.001 max.	0.001 max.
zinc	remainder	remainder
Tensile strength (N/mm^2)	280	330
Elongation (%)	15	9
Solidification shrinkage (mm/m)	0.117	0.117
Relative density	6.7	6.7

to 7.5×10^{-4} mm/mm after five years. Alloy B shows greater shrinkage, ranging from 7.0×10^{-4} mm/mm at five weeks to 14.0×10^{-4} mm/mm five years later. For the majority of purposes, these very small changes in dimensions can be ignored. Where a high degree of dimensional stability is essential, the castings can be *stabilised* by heating them at 100°C for six hours.

Steels and cast iron

The iron–carbon alloys can be divided into two main groups: steels and cast irons. Within each group the carbon content can be varied to give different properties. Steels contain up to 1.2% carbon, while cast irons are normally within the range 2.3% to 4.2%. Alloys with more than 4.6% carbon are rarely used, as they have poor properties.

Steels are not easy to cast but, given the correct techniques, especially in relation to feeder design, castings of high quality can be consistently produced. It is usual to restrict the carbon content to 0.45%, as steels with higher levels tend to give unsound castings which often contain cracks. The properties of suitable steels are discussed on page 147.

Alloys in the iron–carbon system undergo structural changes which are complex but very important in engineering since they control the properties of steels and cast irons. We will be considering these changes in some depth in chapters 10 and 11, but for the present we will concentrate on that part of the equilibrium diagram (fig. 8.10) which covers the cast irons, i.e. 2.3% to 4.2% carbon.

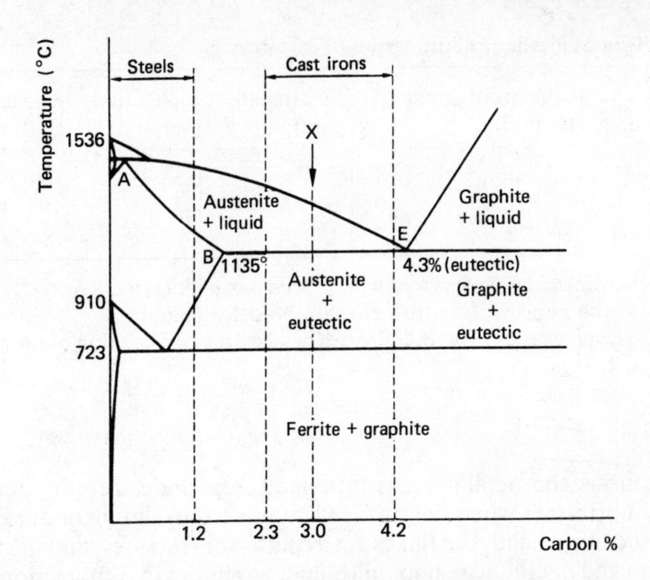

Fig. 8.10 Iron–carbon equilibrium diagram, showing the formation of graphite

The diagram shows that there is a eutectic at 4.3% (point E) which has a solidification temperature of 1135°C. Alloys within the range we are considering (2.3% to 4.2% carbon) are sometimes referred to as hypo-eutectic irons, because their carbon content is below the eutectic composition. Although the melting points of cast irons are much higher than those of most of the non-ferrous alloys we have considered, they are still within the reach of relatively simple melting furnaces. This is one of the reasons for the popularity of iron castings.

When a typical alloy containing say 3% carbon (X in fig. 8.10) is cooled from the molten state, the first material to solidify is austenite, which is a solution of carbon in f.c.c. iron (see section 10.3). The amount of carbon in solution in austenite during solidification is given by the line AB and is less than the average composition of the melt, so carbon is rejected at the boundaries of the austenite to enrich the liquid. The last liquid to solidify has the eutectic composition (4.3%). The eutectic is composed of austenite and carbon. The form of the carbon is dictated by both the cooling rate and the composition, and has an important bearing on the properties of the casting.

If the iron is cooled slowly, that is under near-equilibrium conditions, the carbon occurs as flakes of graphite in a matrix of austenite. On further cooling, the austenite transforms to another phase, called *ferrite*, which is very soft almost pure b.c.c. iron. Irons cooled in this way are grouped together under the general title *grey irons* and are characterised by their good machinability and wear resistance, both of which are attributable to the microstructure of graphite embedded in soft ferrite. The graphite

Table 8.3 Typical mechanical properties of cast irons

Type of iron	0.2% proof stress (N/mm^2)	Tensile strength (N/mm^2)	Elongation (%)	Hardness (HB)
Grey	—	190	0.5	160
White	—	250	—	550
S.G.	180–275	380–550	15–5	180–260

Notes: 1. Hardnesses are measured with a Brinell ball indentor.
2. White irons fracture with virtually no deformation.
3. The properties of an S.G. iron depend on the composition of the melt.

flakes encourage the metal to break into small chips under a cutting action but act as lubricators when the surface is subjected to sliding or abrasive wear. On the other hand, the flakes act as paths for fractures, and the tensile strength and ductility are poor, although properties in compression are good (Table 8.3).

With fast cooling, the carbon is present as iron carbide (cementite) which is a compound with the formula Fe_3C. This is very hard, and the resulting casting is virtually unmachinable but has excellent wear resistance. These *white-iron* castings must be produced to the finished dimensions, as only a light grinding is possible. The fluidity of the molten iron means that thin sections can be readily cast, and the addition of phosphorus (up to 1.2%) further improves the fluidity and enables surface detail to be reproduced with good definition.

Composition also plays an important role in deciding if the carbon is to be in the form of graphite or iron carbide. Additions of silicon encourage the formation of graphite; hence grey irons have between 1.8% and 3.5% silicon, while white irons contain less than 1.2% of this element. Small amounts of magnesium or cerium added to the melt just before casting cause the graphite to grow not as flakes but as small spheroids or nodules uniformly distributed through the matrix. These *spheroidal graphite (S.G.) irons* – which are also called nodular irons – have all the desirable properties of grey irons and in addition have good tensile and impact properties.

Sulphur, on the other hand, prevents the formation of graphite and hence encourages carbide. Sulphur is usually present as an impurity, and its effect can be offset by adding up to 1.5% manganese which combines with it. Manganese sulphide forms small globules which may float to the surface of the melt or be trapped in the casting. If the latter happens, the manganese sulphide does not significantly alter the mechanical properties of the iron and can be regarded as an acceptable inclusion.

Castings in iron and steel are often heat-treated to improve their mechanical properties. Details of these treatments are given in chapter 11.

8.5 Choosing an alloy for casting

It is not easy to choose the most suitable alloy for casting. Not only should the service requirements be considered, we must also evaluate the ease with which the metal can be cast, the technique to be used, and the quality of the finished product. The factors taken into account depend on the application.

A good example is provided by BSI Draft for Development DD 38:1974, which discusses suitable alloys for the manufacture of pump bodies. The assessment in this application is made under the following headings:

a) *Applications* Which parts of the pump are to be cast? Do we require pressure tightness, corrosion resistance, or good properties at elevated or low temperatures?
b) *Heat treatment* Is it necessary? What heat treatments can be applied?
c) *Castability* Are there any limitations on method? What is the thinnest section which can be cast? Are there likely to be problems such as hot cracking?
d) *Weldability* Can surface defects be repaired by welding? Can the component be welded into a larger assembly?
e) *Machinability* What is the optimum speed of machining? Can a good surface finish be achieved?
f) *Properties* What mechanical properties and physical properties (such as thermal conductivity and coefficient of expansion) will the casting have?

Table 8.4 lists the data which could be used in assessing the suitability of phosphor bronze.

Table 8.4 Phosphor-bronze castings for pump bodies (from BSI Draft for Development DD 38:1974)

Data for phosphor bronze. BS 1400:1973–PB1

Description Phosphor bronzes are copper-base alloys containing tin and phosphorus. The structure is made up of three phases:

a) a matrix of copper with tin in solid solution, known as the *alpha* phase, which is comparatively soft;
b) a tin-rich *delta* phase which is hard and is interspersed throughout the matrix;
c) a hard constituent of copper phosphide associated with the *delta* constituent which is also hard but brittle.

PB1 is a very high-grade phosphor bronze, free from zinc and with a low lead content. The structure is ideal for bearing purposes.

Applications Alloy PB1 is recommended for all bearing applications where lubrication is adequate and where toughness, wear resistance, and hardness are required. It has good corrosion resistance to tap waters, mine waters, mild alkalies, and petroleum derivatives.

Castability It is not easy to produce pressure-type castings in this alloy. However, excellent castings are achieved using chill, continuous, and centrifugal casting methods.

Heat treatment Heat treatment is not applied.

Weldability Satisfactory results can be achieved using arc welding (MIG). The alloy can also be brazed and soft-soldered satisfactorily.

Machinability On the basis of comparison between the copper alloys, rather than other metals, machinability of PB1 rates good.

Properties

Chemical composition Chemical composition is as follows:

Element	%min.	%max.	Element	%min.	%max.
Copper	Remainder		Nickel	—	0.10
Tin	10.0	—	Iron	—	0.10
Zinc	—	0.05	Aluminium	—	0.01
Lead	—	0.25	Silicon	—	0.02
Phosphorus	0.50	—			

Mechanical properties (sand castings) at room temperature Mechanical properties at room temperature are as follows:

Mechanical property	Alloy PB1
Tensile strength, min., N/mm^2	220
0.2% proof stress, N/mm^2	130
Elongation, min., %	3
Hardness, HB	70–100

Physical properties Physical properties are as follows:

Physical property	Alloy PB1
Modulus of elasticity, $N/mm^2 \times 10^3$	83–97
Relative density	8.8
Specific heat, kJ/(kg °C)	0.376
Coefficient of thermal expansion per °C $\times 10^{-6}$ (0 °C to 250 °C)	18.3
Thermal conductivity W/(m °C) (15 °C to 200 °C)	47–59
Relative magnetic permeability (H = 8 kA/m)	1.002

9 Metals for forming

Although machining and casting are two of the foremost methods of shaping a metal, forming techniques provide attractive alternatives, especially in terms of mechanical properties of the finished product.

In its broadest sense, the term *forming* covers any method of using plastic deformation to change the shape of a piece of metal. From the point of view of production engineering, it is convenient to distinguish between the processes used on the one hand by the metal producer to make plates, sheets, and sections and, on the other hand, by the manufacturer who shapes these into useful articles. We can call these primary and secondary working respectively.

A metal which has been plastically deformed is said to have been mechanically worked.

9.1 Primary working

The raw material for manufacturing processes is purchased in the form of plate, sheet, bar, rod, or section. Production of these forms starts with an ingot, which is a large casting with a square, round, or hexagonal cross-section. The size and shape of the ingot depends on the type of metal and the method of working used to change it into the required shape. There are three principal methods of primary working: rolling, extrusion, and forging.

Rolling

In the rolling process (fig. 9.1(a)), deformation is achieved by squeezing the ingot between two cylindrical steel rolls which serve two purposes. Firstly, they apply force to the surface of the ingot, which is thus put into compression. Secondly, as the rolls rotate, the ingot is dragged into the gap between them. When the metal passes through the rolls, its thickness is reduced to match the distance between the roll faces. At the same time, it gets longer and wider.

It would be impossible to reduce an ingot, which may be 1 m square, to a 12 mm thick plate in one pass between the rolls: the metal must be passed through many times to achieve the desired reduction. The task is made appreciably easier by heating the ingot. As the temperature of a metal is increased, the yield stress falls and ductility increases significantly. This means that greater reductions are possible at each pass, and less force is needed.

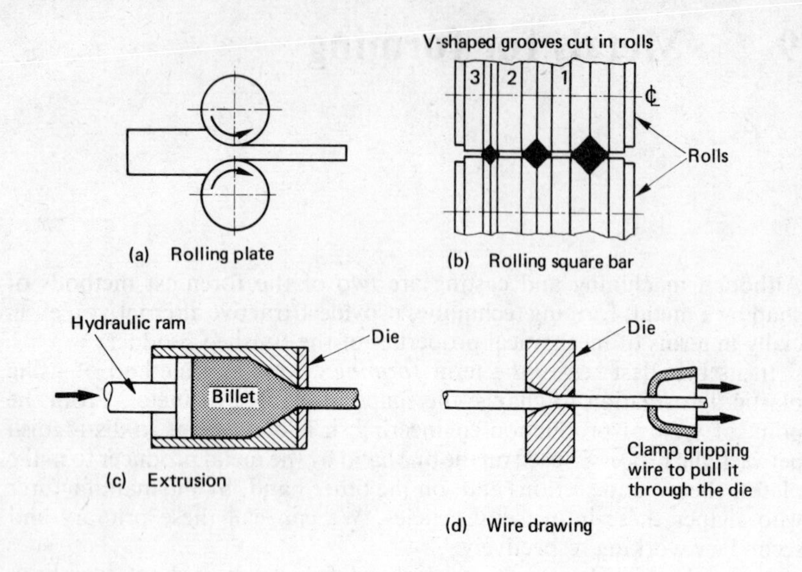

Fig. 9.1 Primary working processes

When the plates have been rolled to the finished thickness, they are guillotined to size and may need levelling or straightening to remove any buckles produced during the rolling process. Hot rolling gives a rough oxidised surface which may need further treatment before dispatch.

Cold rolling is used to produce sheet material from hot-rolled plates. The roll forces are higher than those needed in hot rolling, and a number of passes may be required to achieve a relatively small reduction in thickness. The advantages of cold rolling are that the surface has a smooth finish which does not normally need further treatment, and greater dimensional accuracy can be achieved. Thickness can be controlled to better than 0.1 mm with 2 mm sheet.

A further, not so obvious, benefit accrues from cold rolling. When a metal is cold worked, it becomes harder and the tensile strength increases. The extent of the hardening depends on the amount of deformation that has taken place. This is expressed as a percentage of the reduction in thickness, and the effect can be seen in the data for brass sheet shown in Table 9.1. (Brass is an alloy of copper and zinc.) These changes in properties are reflected in the microstructure, which shows that the grains are broken up and elongated in the direction of rolling.

By controlling the amount of cold working given to the finished sheet, the metal producer can offer a range of strengths for the same composition. Increases in strength can, however, only be made at the price of a loss in ductility. Aluminium sheet can be purchased in the soft, quarter-hard, half-hard, or fully hardened condition. The corresponding properties are shown in Table 9.2.

Table 9.1 Effect of cold work on the tensile properties of brass sheet

Reduction in thickness (%)	Tensile strength (N/mm²)	Elongation (%)
0	305	70
10	353	57
20	399	32
30	460	19
40	506	15
50	552	11

Table 9.2 Mechanical properties of aluminium sheet

Condition	Reduction in thickness (%)	Tensile strength (N/mm²)	Elongation (%)
Soft	0	78	47
Quarter hard	15	98	20
Half hard	30	117	10
Hard	60	151	5

There is a limit to the amount of cold working which can be allowed. As the tensile strength increases so does the yield stress, but at a faster rate. This means that the plastic range is reduced and a point is reached at which any more rolling causes fracture of the sheet. At this stage the metal is work-hardened or, more precisely, strain-hardened. If further cold working is needed to reduce the thickness, the ductility must be restored by heat treating the sheet. During this heat treatment the structure is recrystallised, i.e. the grains are re-formed into their original shape and the metal is softened. We will take another look at work-hardening and recrystallisation later in the chapter.

Hot rolling can also be used to produce a variety of shapes other than flat plate and sheet. Grooves are cut in the rolls which both form the shape and prevent any lateral spread of the metal in the rolls (fig. 9.1(b)). This gives good control of dimensional accuracy. Tables 9.3 and 9.4 are extracts from BS 970: part 1:1972 giving the range of sizes and tolerances on dimensions available with hot-rolled steel bar.

Extrusion

Metals which have very good hot ductility can be formed into sections by extrusion through a die (fig. 9.1(c)). The process is based on the same principles as the toothpaste tube, in which the paste is expelled by squeezing the tube. A hot ingot, usually less than 250 mm in diameter, is placed in a chamber and is forced against a die by a hydraulic ram. The die contains a

Table 9.3 Standard sizes for hot-rolled round, square, hexagonal, and flat bar (from BS 970: part 1:1972)

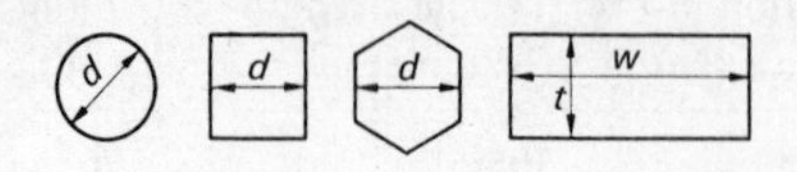

Section	Range	
Round	Diameter, *d*,	$\frac{3}{8}$ in to 10 in (9.5 mm to 254 mm)
Square	Across flats, *d*	$\frac{5}{8}$ in to 6 in (16 mm to 152 mm)
Hexagonal	Across flats, *d*	$\frac{3}{8}$ in to 4 in (9.5 mm to 105 mm)
Flat (narrow)	Width, *w*,	$\frac{3}{8}$ in to 5 in (9.5 mm to 127 mm)
Flat (wide)	Width, *w*,	$5\frac{1}{2}$ in to 18 in (140 mm to 457 mm)

hole which is of the same shape as the required cross-section of the extrusion.

In the main, the process has been confined to metals which can be extruded at relatively low temperatures, e.g. aluminium, magnesium, and copper (fig. 9.2). During the last two decades, however, much progress has been made in developing techniques for the extrusion of steel.

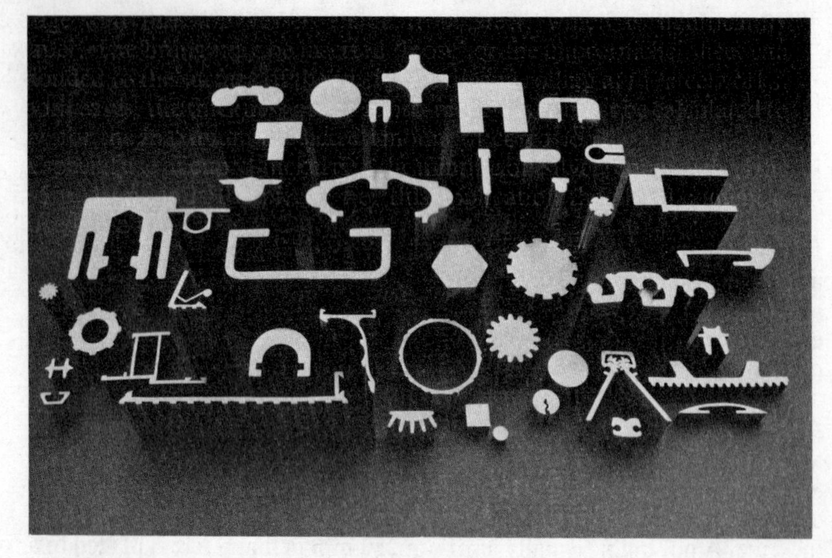

Fig. 9.2 Extruded copper-alloy sections

Table 9.4 Tolerances for general and special applications for hot-rolled round, square, and hexagonal bar under 3 in diameter or width across flats* – inch units (from BS 970: part 1:1972). [Metric values for tolerances are not quoted in BS 970 but may be calculated on the basis 1 in = 25.4 mm.]

	Tolerance†			
	General applications		Special applications	
Size, d [see Table 9.3]	Dia. or width across flats	Out of section‡	Dia. or width across flats	Out of section‡
$\frac{3}{8}$ up to and including $\frac{5}{8}$	± 0.008	0.012	± 0.007	0.011
Over $\frac{5}{8}$ up to and including 1	± 0.012	0.018	± 0.008	0.012
Over 1 up to and including $1\frac{1}{2}$	± 0.016	0.024	± 0.010	0.015
Over $1\frac{1}{2}$ up to and including 2	± 0.020	0.030	± 0.012	0.018
Over 2 up to and including $2\frac{1}{2}$	± 0.024	0.036	± 0.016	0.024
Over $2\frac{1}{2}$ to under 3	± 0.028	0.042	± 0.020	0.030

*These tolerances do not apply to hot-rolled rod.
†By agreement between the purchaser and the supplier, the tolerances may be all plus or all minus. For example, for general application the tolerance for $\frac{5}{8}$ in rounds may be either +0.016 in or −0.016 in.
‡The definition of 'out of section' is as follows:

a) *Round bar* The difference between the maximum and the minimum diameter of the bar measured at the same cross section.
b) *Square bar* The difference between the two dimensions measured across the two pairs of opposing (parallel) sides at a common cross section of the bar.
c) *Hexagonal bar* The difference between the least and the greatest dimensions measured across the three pairs of opposing (parallel) sides at a common cross section of the bar.

Forging

In the forging process, a power-driven hammer is used to reduce the metal to shape. The starting point can be an ingot or a rolled bar. With an ingot, a number of blows are needed to achieve the required amount of deformation. On the other hand, if bar stock is used, the product can often be produced in one blow using metal dies fitted to the faces of the hammer and the anvil (drop forging).

Wire drawing

This (fig. 9.1(d)) is not strictly a primary working process but it has similarities with the techniques described above in so far as volume deformation is involved. Tensile forces are used to draw a rod through a series of dies which progressively reduce its diameter. The drawing is done at room temperature, and work-hardening may occur in some metals before the final diameter is reached. In these cases the wire needs to be heat-treated to restore its ductility before drawing to size.

9.2 Secondary working

Usually, material purchased from the metal producer or stockist needs to be processed further to give the shape required by the customer. Frequently, this means deforming the metal by bending, pressing, deep drawing, spinning, or stretch forming. To a large extent the thickness of the metal determines if it is hot or cold worked. The amount of deformation involved also has a bearing on the working temperature.

In the manufacture of a pressure vessel from 25 mm thick steel plate, the cylindrical shell can be formed by cold rolling. On the other hand, the domed ends are shaped by a spinning operation which involves extensive deformation. If this is carried out at room temperatures, large forces are required and there is a risk of localised work-hardening. For these reasons the ends of a pressure vessel are spun or pressed at a temperature of about 1100°C. By contrast, the domed ends of an air receiver, which are made from much thinner material, can be worked cold because less force is required. Any work-hardening which takes place may even be beneficial, by raising the tensile strength.

9.3 Mechanism of deformation

In both hot and cold working, the prime requirement is that the applied force should exceed the yield stress of the material. This produces deformation which is plastic and is therefore permanent. How does this deformation occur? Why does a material work-harden during cold working but not in hot working? The complete answers to these questions involve complex concepts, but an appreciation of the underlying principles enables us to recognise the role of structure in determining mechanical properties.

In an earlier chapter we likened plastic deformation to cards in a pack slipping over each other. What does this mean at an atomic level? Consider planes of atoms in the close-packed arrangement (fig. 9.3). Neighbouring planes nestle neatly together in fixed positions, but if a shear force is applied they try to move relative to each other. At first, the attractive forces between them keep the atoms in place. If the shear force becomes large enough to overcome the attraction between the planes, however, the atoms in plane A start to move to the left (fig. 9.3(b)). Once they are in line with the atoms in plane B, only a small force is needed to complete the sliding and to bring the planes into register again (fig. 9.3(c)). The planes have now slipped one atomic distance.

On the face of it, this seems to be a neat explanation of slip and we can visualise plastic deformation as slip occurring successively on a large number of planes which move one atomic distance with respect to each other. The problem with this explanation is that it does not take into account the magnitude of the forces which would be involved. To move all the atoms in a plane *at the same time* would require a stress many times higher than the shear stresses observed in mechanical testing. Put another way, in practice we can produce slip at stress levels significantly lower than

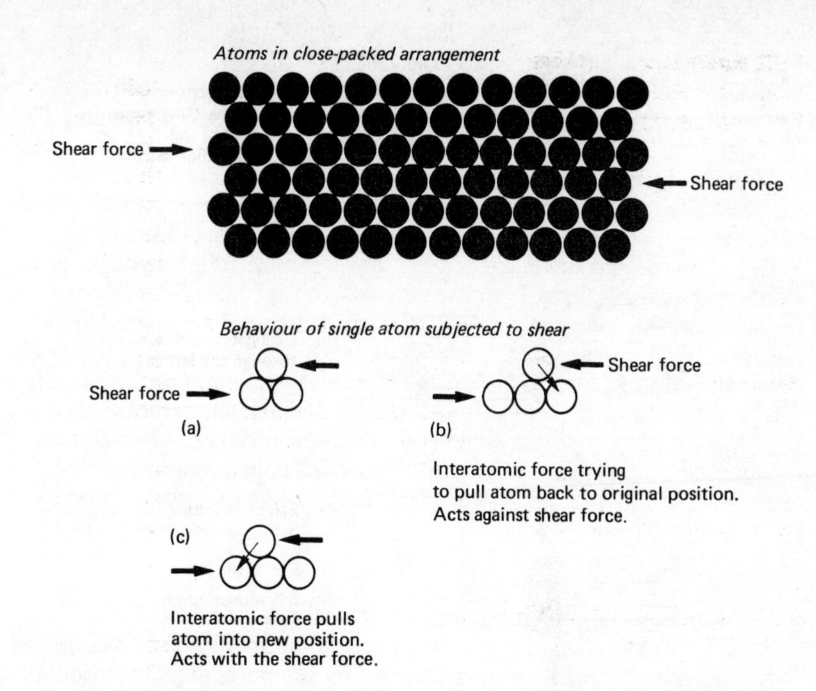

Fig. 9.3 Sliding of atomic planes in shear

the theoretical shear stresses necessary to overcome the attraction between planes. The actual shear stress for a low-carbon steel is about 200 N/mm^2, whereas the theoretical stress needed to shear the planes is about 10 000 N/mm^2.

The explanation of this discrepancy is to be found in the presence of dislocations in the crystal lattice. In chapter 7 a dislocation was identified as an imperfection in the structure, and one type was shown as a mismatch in the atomic planes. This is an edge dislocation and it is formed when the lattice to the left of a vertical line contains one more horizontal plane than the lattice on the right-hand side (or vice versa). A good way of visualising a dislocation is to imagine a thick book, say 1000 pages, in which one page has been cut in half. Most of the pages are unaffected by the presence of the half page and are flat. Those in the immediate vicinity of the half page are bent slightly to accommodate the change in thickness. It is estimated that in an undeformed metal there are about 10^2 to 10^4 dislocations per square millimetre.

At the end of the half plane there is a hole in the lattice, and on each side are atoms which are not in their equilibrium positions (fig. 9.4). The pattern of forces surrounding the dislocations is disturbed, and stresses exist which try to return the system to a stable condition. A shear force acting on the dislocation adds to these and easily moves atom A across the space at the end of the half plane (fig. 9.4(c)). When atom A moves,

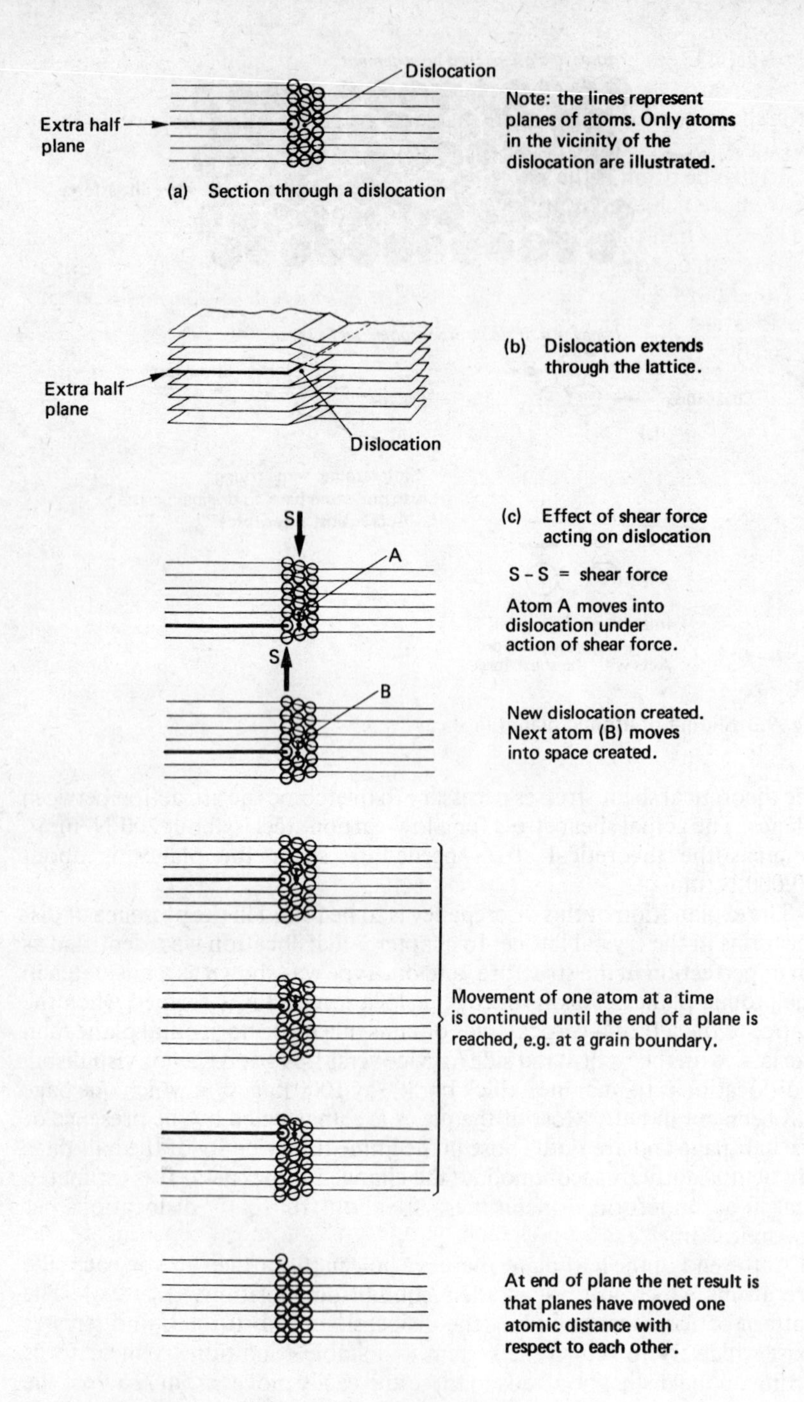

Fig. 9.4 Movement of an edge dislocation

however, it creates a new space in the site it has left. The result is that we have recreated the dislocation one atomic distance away from the original. If the shear stress is maintained, this sequence is repeated until the dislocation has progressed in a series of 'hops' across the grain. When this has occurred, the result is the same as if the whole plane had moved at one go, but we have achieved it at low applied shear stress.

The extra half plane extends through the lattice, of course, and the dislocation can be considered as a tunnel at right angles to the plane of atoms shown in our diagram (fig. 9.4(b)). It is therefore convenient to represent a dislocation as a line which moves across the grain like a piece of taut string (fig. 9.5).

Fig. 9.5 Dislocations in a 0.15% carbon steel (magnified 40 000 times using an electron microscope)

As long as the dislocations can move freely across a grain, only a low applied stress is needed and it is easy to deform the metal. If there is an obstruction in the way, the stress must be increased to push the dislocation past it (fig. 9.6) and we see this as an increase in strength or hardness. A major obstruction is provided by a grain boundary, which acts as a barrier to further movement of the dislocation. It follows that metals which have a large number of small grains are stronger, as the movement of dislocations is restricted by the presence of the grain boundaries and by the short distances that the dislocations can travel before being obstructed. Grain size is thus an important factor in the control of mechanical properties.

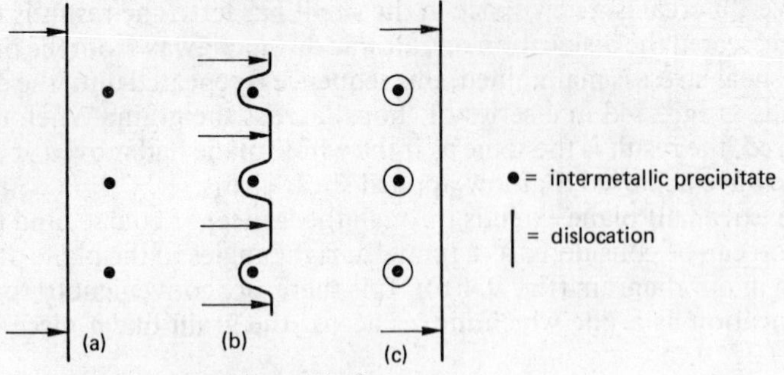

(a) Dislocation moving through lattice.

(b) Dislocation 'stretched' around intermetallic precipitates or globules. Higher stress levels are needed to move the dislocation past the line of precipitates.

(c) Dislocation has moved on, leaving newly formed dislocations in the shape of loops around the precipitates.

Fig. 9.6 Movement of a dislocation through an obstruction

During plastic deformation, new dislocations are generated as existing ones move across the grain. In time, the number of dislocations reaches a level when they start to interact and repel each other, restricting further movement. When this stage is reached the metal has become work-hardened, and attempts to produce further plastic deformation result in fracture. The dislocation density in a work-hardened metal is about 10^{10} per square millimetre.

9.4 Recrystallisation

Although the increased strength of a work-hardened metal is often an advantage, there are occasions when we wish to restore the metal's ductility, e.g. when further working is needed to finish the forming operation. This can be achieved only by reducing the dislocation density.

During plastic deformation, the grains of the metal become distorted and elongated in the direction of the tensile or shear stress (fig. 9.7). If the metal is heated, new grains start to grow at a number of sites or nuclei. These new grains increase their size by capturing atoms from the neighbouring deformed areas. The dislocations are thus eliminated as the lattice in which they existed is restructured. The new lattice has only about the same number of dislocations as the original had before deformation. These re-formed grains continue to grow until they meet and all the distorted material has been absorbed. The structure has now recrystallised. The overall dislocation density is low once again, and the ductility has been restored.

The temperature at which recrystallisation occurs is a characteristic of the metal or alloy. It is not a unique temperature but is affected by the

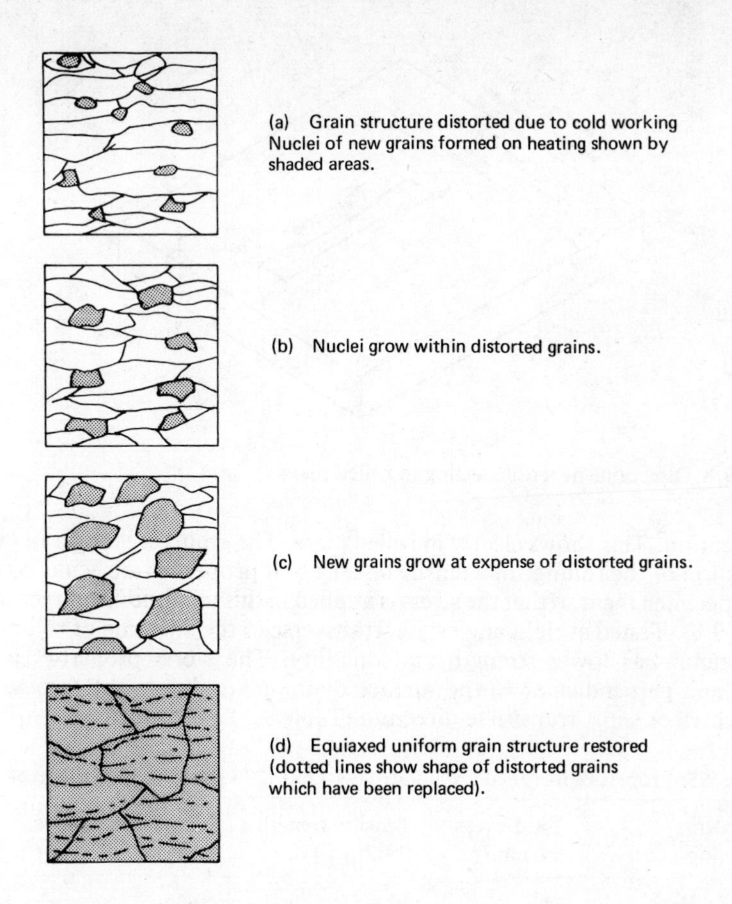

Fig. 9.7 Recrystallisation of a cold-worked metal

amount of plastic deformation which has taken place. Typical temperatures would be 450°C for iron, 200°C for copper, and 150°C for aluminium.

Hot working

We can now see why hardening does not occur during hot working. By keeping the temperature above the recrystallisation point, deformation and growth of new grains proceed simultaneously. A build-up of dislocations cannot occur, because the lattice is being continuously restructured and work-hardening is not possible. The metal retains its ductility, and substantial amounts of deformation can be induced.

9.5 Mechanical properties of worked material

In general, metals which have been worked show directionality in mechanical properties. Even after recrystallisation there is evidence of grain

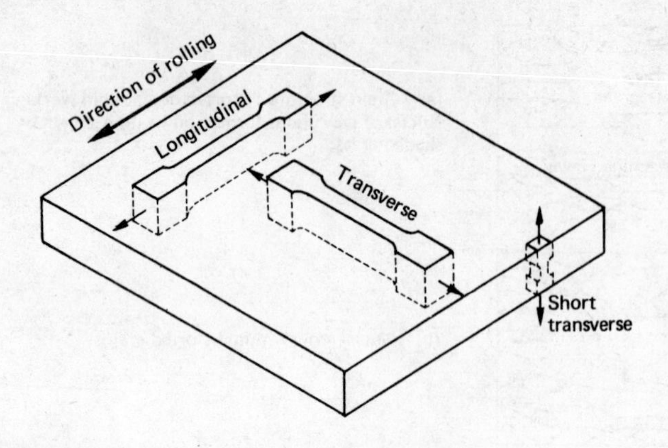

Fig. 9.8 Directions of tensile testing in rolled plate

elongation. This shows clearly in rolled plate. The grains are longer in the direction of the rolling. In a tensile test the best properties are achieved if the specimen is cut so that the stress is applied in this longitudinal direction (fig. 9.8). Tested at right angles, i.e. transverse to the direction of rolling, the metal has lower strength and ductility. The worst properties are obtained perpendicular to the surface of the plate, i.e. in the through-thickness or short-transverse direction (Table 9.5).

Table 9.5 Properties of rolled steel plate (BS 4360:1979 grade 40 – see page 147)

Direction of testing	Yield stress (N/mm^2)	Tensile strength (N/mm^2)	Reduction of area (%)
Longitudinal	280	400	60
Transverse	260	380	60
Short transverse	200	300	15

Hot-worked materials also show a fibre structure (fig. 9.9). This results from the presence of small non-metallic inclusions which are stretched in the direction of working. Fibre lines can provide paths for fracture if subjected to shear along their length. This poses problems in machined components where faces are cut across the fibre lines. Ideally, shear forces should act at right angles to the fibres. Forging a metal enables the shape to be altered while keeping the fibre lines more or less transverse to the shear stress.

9.6 Wrought metals

Metals which have good hot ductility and do not work-harden rapidly when cold worked are ideally suited for the production of worked or wrought products. These metals are generally divided into two groups: ferrous and non-ferrous.

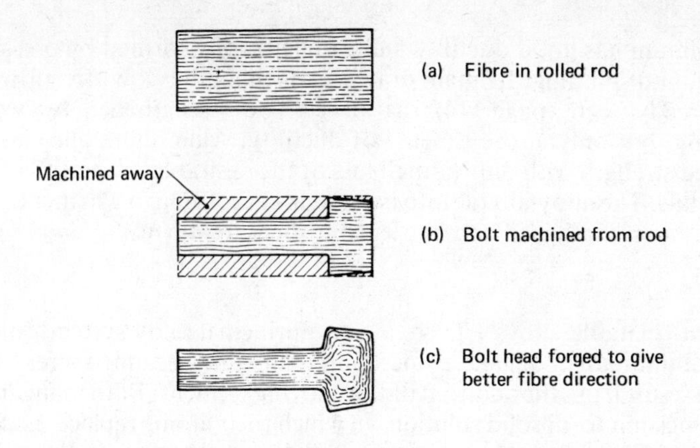

Fig. 9.9 Fibre direction in a bolt head

Ferrous materials are alloys of iron and, as far as we are concerned here, are principally the various grades of steel. They constitute the largest proportion of wrought materials used in engineering, mainly due to their relatively low cost and their versatility. Given the correct choice of composition, steels have excellent formability. The structure and properties of steels are considered in chapters 10 and 11.

The remaining metals are grouped under the general title 'non-ferrous'. In a way this is unfortunate, because it implies a uniformity which does not exist. Non-ferrous metals show a useful diversity. Good strength-to-weight ratios, very high corrosion resistance in severe service environments, excellent creep resistance, and high thermal conductivity are but a few of the properties offered by various non-ferrous metals. In the remainder of this chapter we will be looking at two of the most commonly used groups of wrought alloys to illustrate the range of properties available: aluminium alloys and copper alloys.

Aluminium and its alloys

Aluminium has low density, high electrical conductivity, and good resistance to corrosion. It can be obtained in a very pure condition with less than 0.01% of impurities, i.e. 99.99% purity. Commercial aluminium has small amounts of iron, copper, and silicon present as impurities and can be obtained in purities ranging from 99.5% to 99.8%.

Pure aluminium (99.99%) has the following properties:

Melting point	660°C
Crystal structure	face-centred cubic
Relative density	2.7
Thermal conductivity	240 W/(m °C)
Electrical resistivity	2.66×10^{-8} Ω m
Young's modulus	70.5×10^{3} N/mm^2
Tensile strength	78 N/mm^2

Aluminium has good ductility and can be readily formed by pressing, drawing, and spinning. Its main drawback is that it has low strength. We have already seen (page 114) that it can be strengthened by work-hardening, but only at the expense of ductility. Aluminium alloys offer improved strength with only a small loss of the desirable properties of the pure metals. The alloys divide into two groups according to whether or not they can be strengthened and hardened by heat treatment.

Non-heat-treatable alloys There are two principal alloy systems in this group: aluminium–manganese and aluminium–magnesium. Increases in strength result from the addition of the alloying element. Both manganese and magnesium form solid solutions in which their atoms replace those of aluminium. The lattice becomes strained in the vicinity of the solute atoms, and the movement of dislocations is hindered. The amount of increase in strength depends on the alloy content (Table 9.6).

Table 9.6 Effect of alloy content on strength of aluminium alloys

Alloy addition	Tensile strength (N/mm²)	Elongation (%)
None	78	47
1.25% manganese	110	23
2% magnesium	180	18
3.5% mangnesium	245	14
4.5% magnesium plus 0.75% manganese	330	14

Further improvements in strength can be achieved by cold working during rolling. Some limit is imposed by the more rapid work-hardening which accompanies increases in alloy content. Although it is possible to obtain the 1.25% manganese alloy in the fully hard condition, the ductility of the 4.5% magnesium–0.75% manganese alloy falls to 4% after only 30% reduction in thickness (i.e. the half-hard condition).

The strengths available in commercial alloy plate and sheet are shown in Table 9.7, which is based on BS 1470:1972. It must be remembered, however, that the metal recrystallises if the work-hardened alloys are heated during processing. The properties return to those of the soft condition and cannot be restored except by further working.

The link between alloy content and rate of work-hardening must also be considered in terms of its effect on manufacture. If the material is to be subjected to appreciable amounts of cold forming, it may be necessary to limit the alloy content to avoid work-hardening before the final shape has been produced. The alternative is to accept the cost of heat treatment to restore ductility at some intermediate stage.

Table 9.7 Composition and properties of commercially available aluminium alloys (based on BS 1470:1972)

Type	Designation		Composition (%)				Condition	0.2% proof stress (N/mm²)	Tensile strength (N/mm²)	Elongation (%)
	New	Old	Silicon	Copper	Magnesium	Manganese				
Non-heat-treatable	3103	N3	—	—	—	0.9–1.5	Annealed	—	90	20
							Hardened*	—	160	2
	5083	N8	—	—	4.0–4.9	0.4–1.0	Annealed	125	275	12
							Hardened*	270	345	4
	5154A	N5	—	—	3.1–3.9	—	Annealed	85	215	12
							Hardened*	225	275	4
	5251	N4	—	—	1.7–2.4	—	Annealed	60	160	18
							Hardened*	175	225	3
Heat-treatable	2014A	H15	0.5–0.9	3.9–5.0	0.2–1.8	0.4–1.2	Naturally aged	245	385	13
							Precipation hardened	380	440	6
	6082	H30	0.7–1.3	—	0.6–1.2	0.4–1.0	Precipation hardened	240	295	8

* Corresponds to maximum amount of work-hardening applied to commercially available sheet and plate.

Non-heat-treatable alloys are available in the form of plate, sheet, tubes, and extruded sections. They are used for panelling, lightweight scaffolding tubes, small-boat construction, superstructures for large ships, pressure vessels, and storage tanks. A notable feature of these alloys is that they can be readily welded.

Heat-treatable alloys A more satisfactory way of achieving strength is to heat treat the alloy. As we have seen, not all aluminium alloys respond to heat treatment but there are three main types which do harden – see Table 9.8.

Table 9.8 Heat-treatable aluminium alloys

Alloy additions	Principal uses
1% magnesium + 1% silicon	Architectural fittings (e.g. window frames); structural frames for transport vehicles and containers
4% copper (plus small additions of magnesium, manganese, and silicon)	Highly stressed panels and sections in aircraft. (Often referred to as 'Duralumin', which is a trade name.)
4% zinc + 2% magnesium	Structural components such as light bridges and television masts; armour plate in fighting vehicles

The heat treatment which is applied to these alloys can be either age hardening or precipitation hardening. The key to the success of these treatments lies in the fact that all the alloys mentioned have limited solid solubility which decreases as the temperature falls. We can see how this enables us to strengthen the metal by considering what happens to an aluminium–4% copper alloy (fig. 9.10).

We start by heating the alloy to 500°C. All the copper is in solid solution at this temperature (α phase). When the metal is cooled, copper is progressively rejected from the solution until it has been reduced to 0.2%, which is the solubility limit at 20°C. The copper which is released from solution forms an intermetallic compound $CuAl_2$. Under equilibrium conditions, therefore, the microstructure of the alloy is grains of α phase with islands of $CuAl_2$ at the grain boundaries. In this condition the metal has a hardness of about 65 HV, due to the intermetallic compound being hard. For comparison, the hardness of pure aluminium is 22 HV.

On the other hand, if the alloy is quenched in water from 500°C, insufficient time is allowed for the copper to come out of solution. It is retained in the lattice in a non-equilibrium condition. The lattice is strained and tries to reject the copper. At room temperature, copper diffuses only very

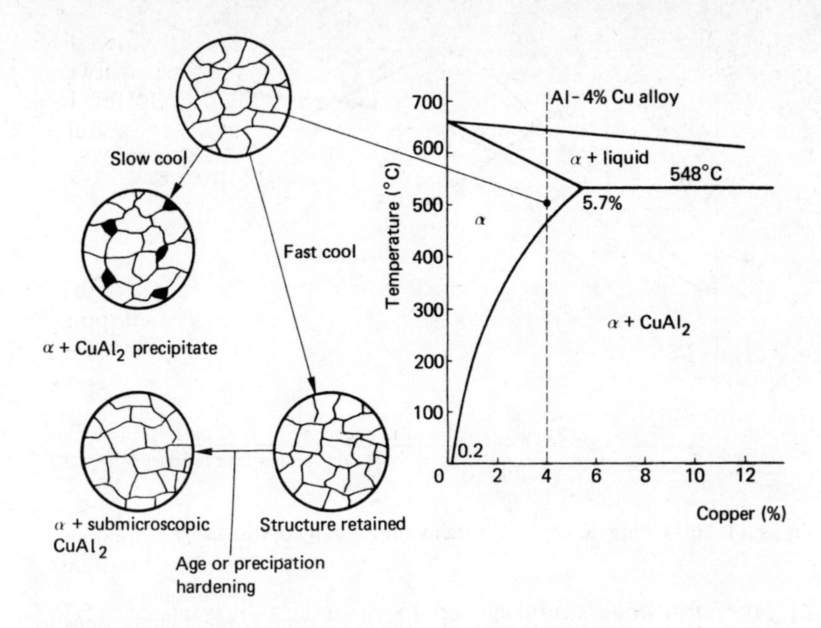

Fig. 9.10 Precipitation hardening in aluminium–copper alloys

slowly. It takes three or more days for the copper atoms to move from the lattice and to collect in small clusters where they combine with aluminium to form very small particles or precipitates of $CuAl_2$. These are so small that they cannot be seen under a metallurgical microscope but they do provide effective barriers to the movement of dislocations and so harden the alloy. The hardness increases with time and after seven to ten days reaches about 120 HV. Further increases of a few points in hardness are observed as time goes on.

The precipitation process can be speeded-up by reheating the alloy, after quenching, to between 120°C and 180°C. High levels of hardness (135 HV) can be achieved in a matter of hours (fig. 9.11). After prolonged heating, the hardness falls, i.e. over-ageing occurs. This is associated with the movement of the submicroscopic $CuAl_2$ precipitates to form large globules which can be seen under the microscope. These are too large and too far apart to hinder the dislocations, and the material becomes soft. The time required to achieve optimum hardness varies from 15 hours at 120°C to 3 hours at 180°C. The precipitation temperature is chosen to give the best compromise between mechanical properties and production requirements.

Aluminium–4% copper alloys are capable of giving tensile strengths of 400 to 500 N/mm^2 with an elongation of 10% after precipitation hardening.

To summarise, there are two heat treatments which can be used to harden suitable aluminium alloys:

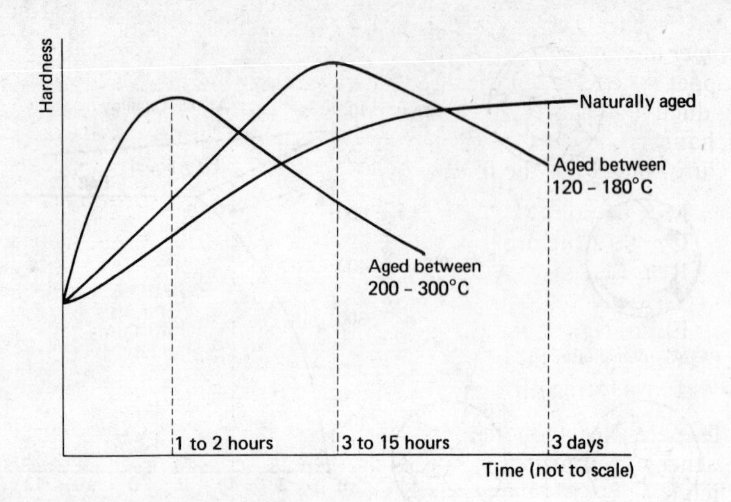

Fig. 9.11 Hardening and over-ageing in aluminium copper alloys

a) *Age hardening* (natural ageing), in which the alloy is heated to 500°C, quenched, and kept at room temperature while ageing takes place. This can be applied to both the aluminium–copper and aluminium–zinc–magnesium alloys, but not to the aluminium–magnesium–silicon group.
b) *Precipitation hardening*, in which the quenched alloy is reheated to give controlled precipitation of an intermetallic compound. This is suitable for all three groups of alloys.

The one remaining question to be answered is why we cannot heat treat other alloys which have limited solubility, such as the aluminium–magnesium group. The reason lies in the form of the precipitate. The non-heat-treatable alloys have large globular precipitates which do not affect the movement of dislocations through the lattice and thus do not increase the amount of effort required to deform the metal, which is the same as saying that the hardness is not increased.

A particular advantage of using a heat-treatable alloy in fabrication is that it can be cold worked in the quenched condition, when it has good ductility. Subsequent reheating to 500°C, as part of the hardening process, not only takes all the alloying element into solution but also recrystallises the structure. Ageing or precipitation hardening is then carried out on the finished article.

Unlike the non-heat-treatable alloys, which are readily weldable, only low-strength joints can be produced in aluminium–copper and aluminium–magnesium–silicon alloys. Good-quality high-strength welds can be achieved, however, in aluminium–zinc–magnesium alloys.

Copper and its alloys

Copper is best known for three outstanding properties: its high electrical conductivity (cables and wires), thermal conductivity (pipes and heat exchangers), and corrosion resistance (chemical and brewing plant).

Pure copper has the following properties:

Melting point	1083°C
Crystal structure	face-centred cubic
Relative density	8.93
Thermal conductivity	393 W/(m °C)
Electrical resistivity	1.67×10^{-8} Ω m
Young's modulus	122.5×10^3 N/mm^2
Tensile strength	230 N/mm^2

The electrical conductivity of copper is markedly reduced by the presence of impurities. A special grade, *high-conductivity (HC) copper*, which has a very low impurity content, is produced for electrical work.

For most other purposes copper can contain small amounts of impurities without detriment. An exception to this is oxygen, which is undesirable if copper has to be welded, since it gives porosity in the weld metal. As produced, copper normally contains a significant amount of oxygen. In this condition, it is referred to as *tough pitch copper*. The oxygen can be removed by adding a deoxidant to the molten metal during manufacture. Phosphorus is widely used to deoxidise copper. There is always a small amount of phosphorus left after deoxidation. As little as 0.04% residual phosphorus reduces the electrical conductivity to about 75% of that of high-conductivity copper. *Deoxidised copper* is used for any product which is to be welded, e.g. tubes, storage tanks, heat exchangers, and brewery plant. *Arsenical copper*, containing up to 0.5% arsenic, is also used for welded plant where resistance to scaling is required, e.g. direct-fired boilers.

Aluminium, nickel, tin, and zinc readily alloy with copper. As small additions, they go into solid solution and enhance the strength. With high alloy contents, intermetallic compounds are formed and the equilibrium diagrams are complex.

The bronzes (copper–tin alloys) were discussed in chapter 8, as they are principally casting alloys. The three most common groups of wrought alloys are the brasses, aluminium bronzes, and cupro-nickels.

Brasses are alloys of copper and zinc. Two compositions are used for engineering work: 70% copper–30% zinc and 60% copper–40% zinc (Muntz metal). Both the 70/30 and 60/40 brasses have good corrosion resistance, especially in marine environments, and in the annealed condition they offer strengths of 325 to 375 N/mm^2. They both have good machinability, and to a great extent the choice between the two is based on forming requirements. 70/30 brass has very good ductility (65% to 75% for annealed sheet) and does not work-harden rapidly. It is thus well suited to deep drawing, pressing, and spinning. Although 60/40 brass tends to work-harden more rapidly, it shows better hot-working properties and is to be preferred for extrusions, forging, and hot stamping.

Aluminium bronzes are copper–aluminium alloys – they do not contain tin, even though they are called bronzes.

There are two principal alloys. 95% copper–5% aluminium has a tensile strength of 400 N/mm^2 which can be increased to nearly 700 N/mm^2 by cold rolling. The strength of the second alloy, 90% copper–10% aluminium, depends on the amount of iron (up to 2.5%) and nickel (2% to 5%) which has been added. This alloy can also be heat-treated to improve the strength, but at the price of low ductility. In general, tensile strengths within the range 550 to 750 N/mm^2 are readily attainable.

The principal attractions of aluminium bronzes are their high resistance to corrosion and wear, good fatigue strength, and fine surface finish. They are used in petrochemical and other processing plant, especially as valve and pump bodies. They are also used in imitation jewellery, due to their bright gold surface lustre.

Cupro-nickel is, in the main, a 70% copper–30% nickel alloy. On page 88 we saw that this alloy is a solid solution of nickel in copper. As with other solid solutions based on copper, it has good cold-working properties and can be readily formed.

As it has very good corrosion resistance, cupro-nickel is also used in chemical plant and is often referred to by its trade name 'Monel'. With a tensile strength of 375 N/mm^2, it can be used as a load-bearing component in pressurised systems. It is an expensive alloy, however, and in the interests of economy it is often used simply as a corrosion-resistant lining for storage and pressure vessels made from steel, which is significantly cheaper. The lining may be inserted 'loose' or the cupro-nickel sheet can be bonded to the surface of the steel plate during rolling at the steelworks. In both cases, the thickness of the cupro-nickel layer must be calculated to allow for the maximum predicted amount of corrosion.

A special use of cupro-nickel is in the minting of 'silver' coins such as the UK 5 p, 10 p, and 50 p pieces. For this application the nickel content is reduced to about 25%.

10 Structure and properties of steels

Of all the factors which contributed to the success of the industrial revolution in the last century, one of the most important must surely have been the availability of steel in commercial quantities. There is no record in history of the actual discovery of steel, but swords and knives made from this metal have been valued for centuries because of their sharp cutting edges which did not blunt rapidly in use. These same attributes were essential for the tools which were needed in the industrial revolution to machine drive shafts, gears, and other engineering components. The growth of industry and the transport system also created a need for a more adaptable material than cast iron. Designers looked for a strong ductile metal which they could use in structures such as ships, trains, bridges, and boilers, and found that low-carbon steel (originally known as mild steel) satisfied their requirements. In addition, methods developed by Bessemer and Siemens enabled this steel to be manufactured in large quantities at relatively low prices.

Steel is so commonplace in our everyday lives that we often overlook the fact that there are many types, each designed by the metallurgist to have some particular property required by the engineer or designer. The requirements for a steel vary considerably but more often than not involve consideration of tensile properties and/or impact strength. Some examples of typical applications with their associated requirements are listed in Table 10.1.

If we look at the compositions of the steels which meet these requirements, we observe that they all contain carbon, manganese, silicon, phosphorus, and sulphur. The last two elements are impurities introduced by the raw materials used to make the steel. The steelmaker tries to keep these as low as possible consistent with economic production. Normally the sulphur and phosphorus contents do not exceed 0.05% and we do not need to take them into our consideration. The small amounts of silicon are residues from the steelmaking process and do not directly influence the properties of the steels listed in Table 10.1. We are left, therefore, with carbon and manganese. Both of these alloying elements have a significant influence on the mechanical properties of the steel, but the more important is the carbon. The amounts of carbon in steel are small when compared with the alloy contents of the non-ferrous metals which we discussed in chapter 9. Nevertheless, even small variations in the carbon content have a profound effect not only on the properties of the steel but also on the way in which these can be altered by heat treatment.

Table 10.1 Typical carbon steels

Application	Requirements	Composition (%) C	Mn	Si	P	S
Car body panels	Sheet which can be pressed accurately into shape, especially around sharp corners etc. A suitable steel must have good ductility with a reasonably low yield stress.	0.08	0.3	—	0.05	0.05
Ships' hulls	Plates must have good strength, to withstand the stresses experienced in service in heavy seas. There is also a need for good impact strength at sub-zero temperatures, to avoid catastrophic brittle fractures. The steel used must also be weldable, since all modern ships' hulls are fabricated by welding.	0.18	0.8	0.1	0.05	0.05
Axle shafts	The bar stock used for this application must have good strength in bending and torsion. It is also often desirable to harden the surface layers to improve resistance to wear.	0.4	0.8	0.1	0.05	0.05
Helical springs	The steel must first be capable of being rolled into rods. The next requirement is good ductility, to enable the coils to be formed. Finally the steel must respond to heat treatment to give suitable mechanical properties.	1.0	0.6	0.3	0.05	0.05

10.1 Iron–carbon alloys

The first important fact to note about steels is that they are part of a range of iron–carbon alloys which we first met in chapter 8. As we saw there, iron and carbon combine together to form a compound, iron carbide. In this, one carbon atom is bonded to three iron atoms; hence it is given the simple formula Fe_3C, although its crystal structure is complex. When we are talking about the structure of iron–carbon alloys, the carbide is usually called *cementite*.

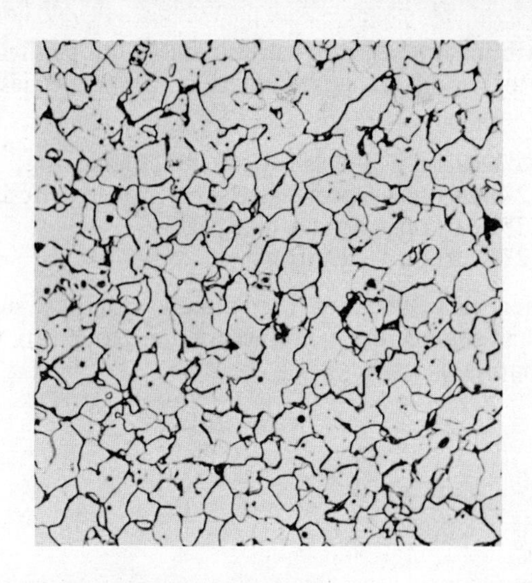

Fig. 10.1 Ferrite grains (magnification $\times 100$)

The amount of cementite present in an iron–carbon alloy depends on the carbon content. If there is little or no carbon present, the microstructure consists of uniform grains of *ferrite* (fig. 10.1) with no cementite and the metal is soft and ductile. Ferrite is the name given to pure body-centred-cubic iron with a very small amount of carbon in solution. At the other end of the scale, with a carbon level of 6.67%, the structure consists entirely of cementite: the material is very hard with virtually no ductility and no commercial use. Between these two extremes, the ductility depends on the proportion of cementite in the microstructure and hence on the carbon content (Table 10.2).

Table 10.2 Ductility of plain carbon steels

Carbon content (%)	Elongation in tensile test (%)
Nil (i.e. pure iron)	42
0.2	37
0.4	31
0.6	22
0.8	17
1.2	3

Note: carbon–manganese steels would give lower elongations than those quoted above (see Tables 10.4 and 10.5).

Based on carbon content, iron–carbon alloys can be divided into four groups which offer different combinations of properties and characteristics:

a) 0.1% to 0.8% carbon – steels for general engineering;
b) 0.9% to 1.2% carbon – special steels, e.g. for wear resistance;
c) 1.3% to 2.2% carbon – not normally used;
d) 2.4% to 4.2% carbon – cast iron.

In this chapter, we will be concerned mainly with those steels which are in common use in engineering manufacture (i.e. group (a)). Cast iron was discussed in chapter 8.

10.2 Strength versus carbon content in steels

The role of carbon can be seen most readily in the range of strengths which is available from various steels (fig. 10.2). As the carbon content is increased from zero to 0.8%, the tensile strength goes from 250 N/mm^2 to 850 N/mm^2. These changes in strength must be related to the microstructure. As we have already seen, at very low carbon contents the metal consists entirely of grains of ferrite. At 0.4% carbon, when the strength is about 540 N/mm^2, just over half the area viewed through a microscope is occupied by a constituent known as pearlite (fig. 10.3). In a steel containing 0.8% carbon, the metal is completely pearlitic and the maximum strength is reached (fig. 10.4).

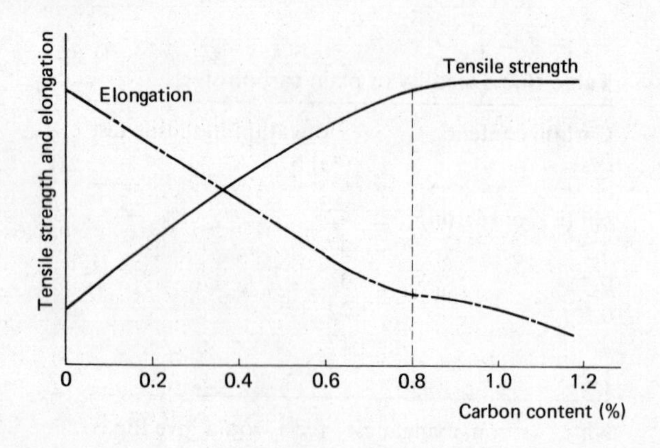

Fig. 10.2 Relationship between strength and carbon content

Fig. 10.3 Microstructure of a 0.4% carbon steel after slow cooling (magnification $\times 150$)

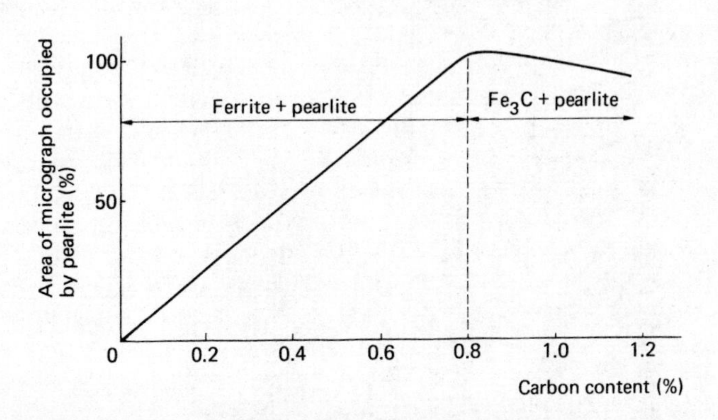

Fig. 10.4 Relationship between pearlite % and carbon content

It follows that there is some relationship between the presence of pearlite and the increase in the strength of the steel. By plotting tensile strength as a function of the percentage of the area occupied by pearlite (fig. 10.5), we can see that the relationship is similar to the one shown in fig. 10.2 between strength and carbon content. The link is to be found in the structure of pearlite. At low magnifications, etched pearlite often appears to be a uniform material with the colours of mother-of-pearl – this is the origin of its name. Examination at higher magnifications (fig. 10.6) shows that it has a laminated structure. It is a composite consisting of alternate layers of

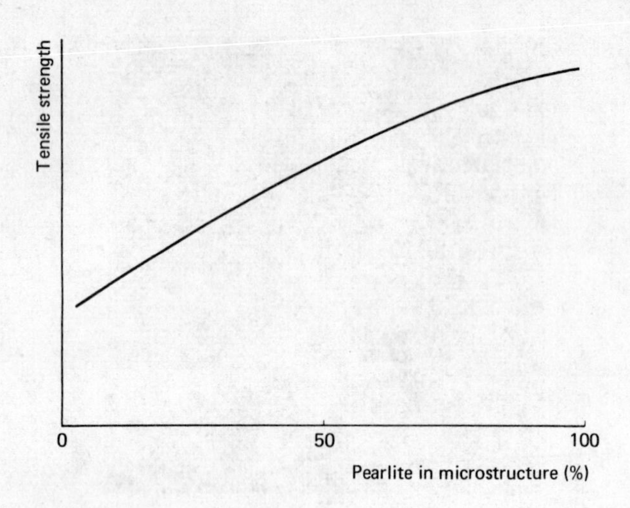

Fig. 10.5 Relationship between pearlite % and strength

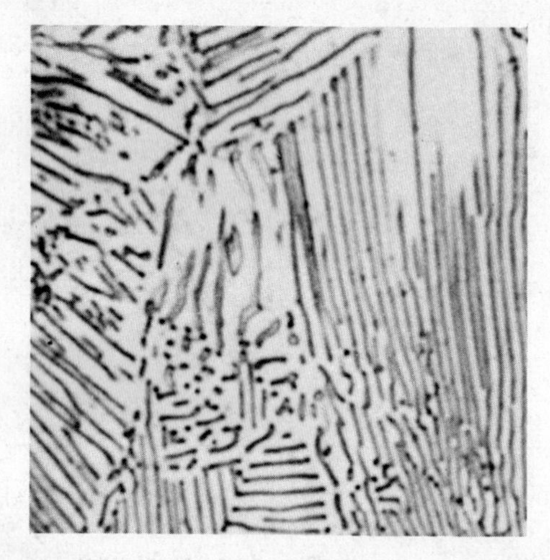

Fig. 10.6 The structure of pearlite in a 0.8% carbon steel (magnification ×1500)

ferrite and cementite. Since the amount of cementite in a steel is determined by the carbon content, it follows that the primary effect of carbon in relation to strength is seen in the amount of pearlite it produces. In other words, increases in carbon content act indirectly through the changes which they cause in the microstructure.

This link between microstructure and properties is an important observation since, as we will see later, heat treatment can be used to change the structure of steel and give different strengths with the same carbon content.

Within the range 0.9% to 1.2% carbon, the steel continues to have a pearlitic structure but cementite forms a network at the grain boundaries. Since it is not an integral component of the pearlite, it is referred to as *free* cementite. This network of a hard brittle compound at the grain boundaries is associated with an increase in the hardness, but it causes a marked deterioration of the ductility and this is undesirable for most commercial purposes.

In general, the strength does not increase with carbon content between 0.9% and 1.2%. At best, it remains that of the fully pearlitic structure, but it may even fall to a value below that of the 0.8% carbon steel.

10.3 Formation of pearlite

As pearlite is so obviously a critical constituent in the structure of a steel, it is important to understand how it is formed and controlled. The formation of pearlite, and indeed the whole metallurgy of steels, relies on the allotropy of iron. A metal is allotropic if it can exist in two or more crystalline forms.

Allotropic forms of iron

Iron has two allotropic forms which have different atomic structures. The first of these, α (*alpha*) iron, has a body-centred-cubic lattice and exists at room temperature. When raised to elevated temperatures, the atoms rearrange themselves and adopt a face-centred-cubic arrangement, γ (*gamma*) iron. On cooling, the structure reverts to body-centred-cubic, the change taking place at 910°C. These two forms of iron have different physical properties, e.g. coefficients of expansion. Energy is expended in the change from γ to α iron, and the cooling curve shows a thermal arrest at the temperature at which the event occurs. The terms *transformation* or *reaction* are often used when referring to this change from one allotropic form to the other.

Tranformation of austenite

During the cooling of a piece of pure iron, the transformation occurs at a unique temperature (910°C). Above this point the metal is all in the γ form (i.e. face-centred-cubic lattice), while below it the entire crystal structure is body-centred cubic (i.e. the α form).

If carbon is added to the iron to produce a steel, the change from γ to α takes place over a range of temperatures. The limits of this range are known as the critical points and, by convention, the higher temperature – i.e. the one at which the transformation starts during cooling – is designated A_3 while the lower point is A_1. The A_1 temperature remains constant at 723°C irrespective of the carbon content. On the other hand, the A_3 temperature is progressively reduced as the carbon content is increased, until it reaches 723°C at 0.8% carbon. Above this level, the A_3 temperature rises again.

Plotting the A_3 and A_1 temperatures for a number of steels with carbon contents up to 1.2% produces an equilibrium diagram which is part of the

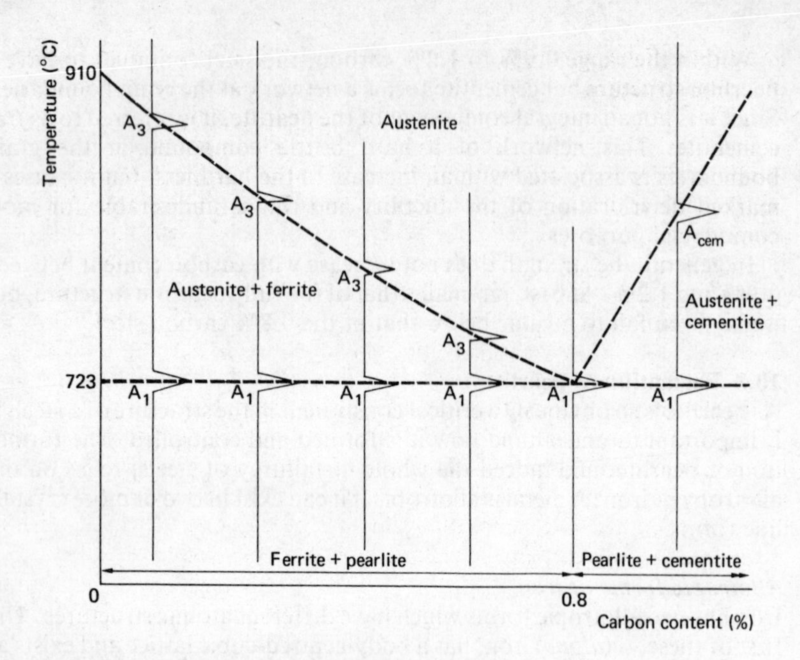

Fig. 10.7 Critical points for carbon steels

larger iron–carbon diagram (fig. 10.7). It shows the transformation under equilibrium cooling conditions of austenite, which is the name given to γ iron with carbon in solution. Although the diagram shows a solid-to-solid transformation, we can apply the rules of interpretation which we used for equilibrium diagrams concerned with solidification.

The first point we need to note is the difference in carbon solubility between austenite and ferrite. The face-centred-cubic lattice can hold as much as 1.7% carbon in solution at 1130°C. Thus, with the range of steels we have been discussing, the structure is entirely austenitic at this temperature and all the carbon is in solution. The carbon is dissolved interstitially: in other words, a carbon atom does not replace an iron atom to form part of the cube but nestles between two iron atoms (fig. 10.8). When the

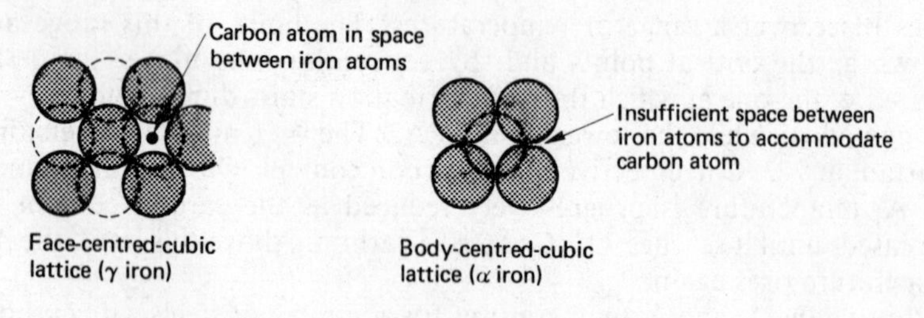

Fig. 10.8 Interstitial solution of carbon

lattice is rearranged to give a body-centred-cubic pattern, there is inadequate space between the iron atoms and the carbon atom is ejected from its site. This is the reaction which takes place at the A_3 point.

Looking at the event diagrammatically, we can visualise the b.c.c. ferrite grain being nucleated at the edge of an austenite grain (fig. 10.9) with the result that a boundary is set up between the two constituents. As the temperature is lowered, the boundary advances further and further into the austenite grain. When the carbon is rejected from the ferrite lattice, it diffuses into the body of the austenite grain and is retained in solution. This means that the untransformed austenite becomes progressively richer in carbon until the amount of this element in solution reaches 0.8%. At this point the events are very similar to those discussed on page 92 in relation to eutectics: the remaining austenite is completely changed to ferrite and virtually all the carbon is rejected from the solution to form layers of cementite which are sandwiched between ferrite. This eutectic-type reaction occurs at the A_1 point, and the product is pearlite. As this takes place in the solid state, the term *eutectoid* is used, to distinguish it from the eutectic structures developed when metals solidify.

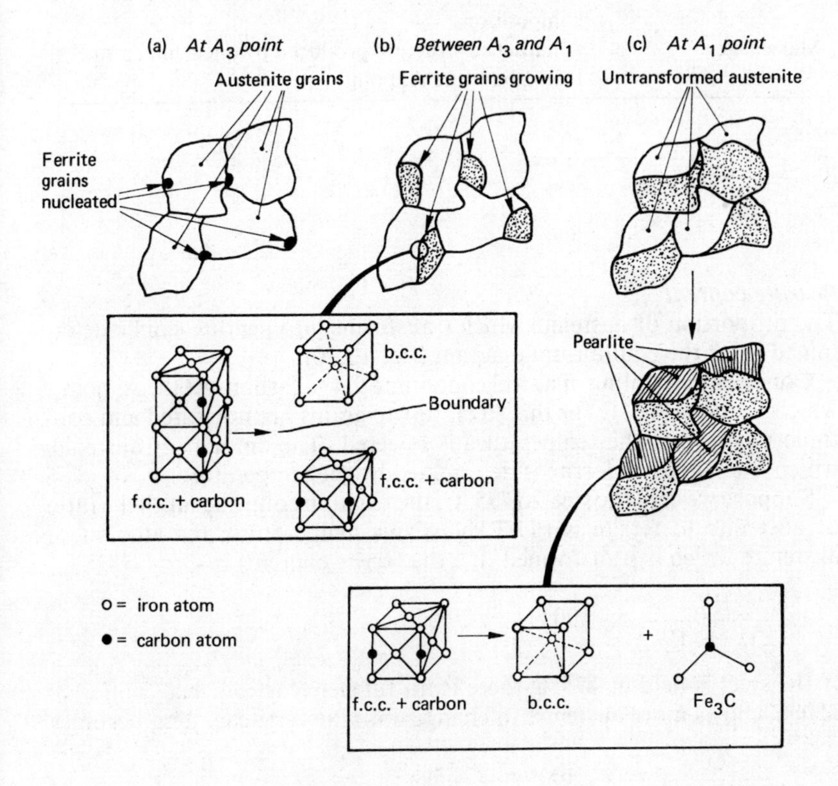

Fig. 10.9 Transformation of austenite

Some of the terms used to describe the structure of steels are summarised in Table 10.3.

Table 10.3 Terms used to describe the structure of steels

Term	Explanation
α iron (*alpha* iron)	Pure iron which has a body-centred-cubic lattice structure
γ iron (*gamma* iron)	Pure iron which has a face-centred-cubic lattice structure
Ferrite	α iron with a small amount of carbon in solution; ferromagnetic
Austenite	γ iron with carbon in solution; non-ferromagnetic
Cementite	A compound of iron and carbon (Fe_3C)
Pearlite	Eutectoid structure composed of alternate layers of ferrite and cementite
Hypo-eutectoid steels	Steels which have a carbon content less than that of the eutectoid (0.8%)
Hyper-eutectoid steels	Steels with a carbon content above the eutectoid composition
Martensite	A hardened structure produced by quenching a steel from above the A_3 point.

Pearlite content

The proportion of austenite which transforms into pearlite can be determined from the equilibrium diagram (fig. 10.10).

Consider the cooling of a steel containing 0.4% carbon. At the A_3 point, which is about 800°C for this steel, ferrite grains are nucleated and continue to grow as the temperature is lowered. The amount of austenite which is changed to ferrite is determined by the temperature.

Suppose we cool the steel to 775°C; the reaction continues until the ratio of austenite to ferrite is PU/PV. At this temperature, the amount of austenite which is transformed (i.e. the ferrite content) is

$$\frac{PV}{PU + PV} \times 100\% = 29\%$$

If the steel is held at 775°C there is no further reaction, but cooling to 750°C allows more austenite to change until the ferrite content becomes

$$\frac{QX}{QW + QX} \times 100\% = 45\%$$

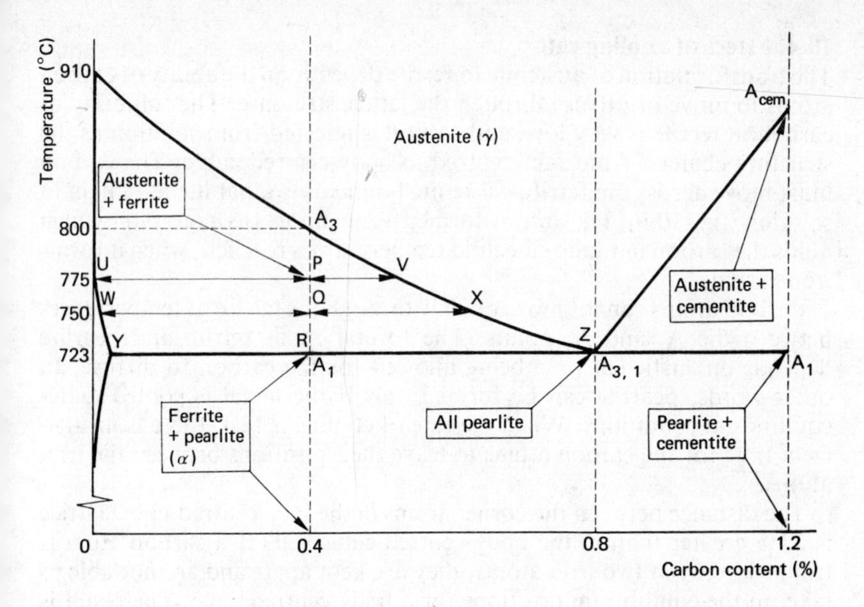

Fig. 10.10 Structure of steels

At 723°C the austenite which is left is transformed entirely into pearlite. This means that the pearlite content of the steel is

$$\frac{RY}{RY + RZ} \times 100\% = 59\%$$

Since the point R is determined by the carbon content – in this case 0.4% – it follows that in any steel the amount of pearlite is controlled by the amount of carbon which is present. The limiting case is at 0.8%, where R coincides with Z and there is 100% pearlite. This confirms the microscopic evidence discussed on page 136.

Cooling a 1.2% carbon steel With this steel, the austenite at 1000°C contains more carbon in solution than the eutectoid composition. At the onset of the transformation, at approximately 875°C, carbon is again rejected from solution but in this case forms free cementite at the grain boundaries. The austenite grain is thus depleted of carbon. For clarity, this temperature is referred to as the A_{cem} point, to distinguish it from the austenite-to-ferrite transformation point (A_3). As the temperature falls, more cementite is deposited at the boundary and the austenite approaches the eutectoid composition. At 723°C the remaining austenite contains 0.8% carbon and transforms to pearlite. The final structure consists of pearlite with a network of cementite at the grain boundaries.

10.4 Effect of cooling rate

The transformation of austenite to ferrite depends on the ability of carbon atoms to move or diffuse through the lattice structure. The solubility of carbon in ferrite is very low, and carbon is rejected from solution as the structure changes from face-centred to body-centred cubic. The carbon must move across the ferrite–austenite boundary so that it can remain in solution (fig. 10.9). The untransformed austenite acts as a reservoir which holds the carbon until the eutectoid temperature is reached, when it forms iron carbide.

Carbon atoms cannot move quickly through the lattice at temperatures between the A_1 and A_3 points. The formation of ferrite and pearlite depends on sufficient time being allowed for the carbon to diffuse. In other words, pearlite can be formed only if the metal is cooled under equilibrium conditions. When the rate of cooling is fast, there is insufficient time for the carbon atoms to leave their positions between the iron atoms.

The distance between the corner atoms in the face-centred-cubic lattice is 23% greater than in the body-centred-cubic cell. If a carbon atom is trapped between two iron atoms, they are kept apart and are not able to take up the equilibrium positions for a body-centred cube. The result is that the lattice is distorted. Dislocations cannot move freely in a distorted lattice, and the steel becomes harder and loses its ductility. This is what happens when a chisel is hardened by heating it to just above the A_3 temperature and then plunging it into water or oil (i.e. quenching). The quenching gives a fast cooling rate and reduces the temperature to about 300°C before the carbon has had time to diffuse. The austenite-to-ferrite transformation is suppressed and the material is permanently hard, since carbon diffusion rates are negligible below 300°C and the carbon atoms stay fixed between the iron atoms. The lattice structure remains distorted and can be rectified only if the temperature is raised again to allow the carbon to diffuse. Under the microscope the structure looks acicular or needle-like and is known as *martensite* (fig. 10.11).

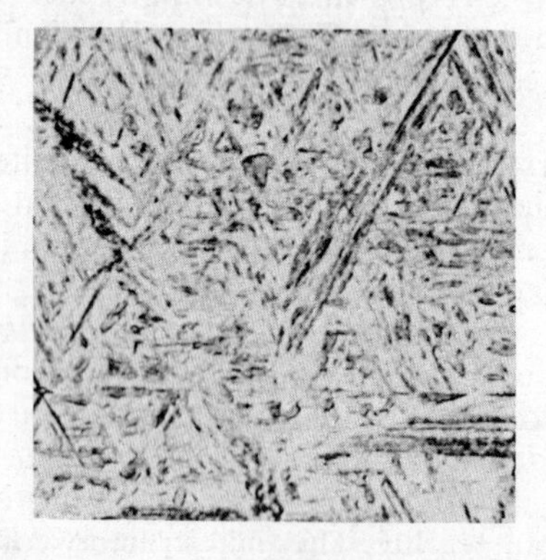

Fig. 10.11 Martensitic structure in a 0.4% carbon steel (magnification ×500)

The increase in hardness achieved by quenching depends on the carbon content. At low carbon levels there is very little change. The effect begins to be more noticeable at about 0.4% carbon and reaches a maximum at the eutectoid composition. The increases at 0.4% and 0.8% carbon are 255 HV and 530 HV respectively.

Since the formation of martensite depends on the cooling rate, we need to know how fast the steel must be cooled to produce hardening. This is the *critical cooling rate* and for carbon steels it is somewhere between 400°C/s and 500°C/s. Carbon content affects the critical cooling rate. As the carbon content increases, the critical rate becomes slower (fig. 10.12). This means that it is easier to harden an 0.8% carbon steel than one containing 0.4%.

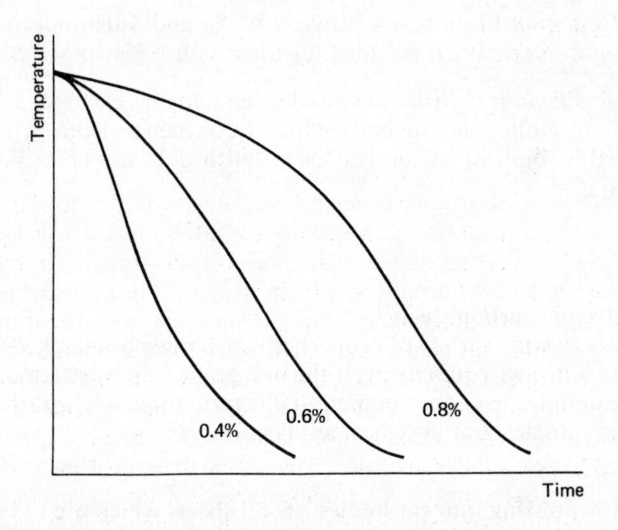

Fig. 10.12 Critical cooling rates for different carbon contents

In normal heat-treatment practice, quenching into water represents about the fastest cooling rate we can achieve. This corresponds to the critical rate for an 0.35% carbon steel. Hence steels with this carbon content or less are not suitable for applications where the component is to be quench-hardened.

10.5 Alloying elements

Frequently alloying elements are added to carbon steels to achieve specific properties, such as higher strength or better corrosion resistance. Sometimes the aim may be to make it easier to harden the steel, especially at lower carbon contents.

There are many alloying elements which can be added, but the following are the four most commonly used:

a) *Manganese* increases strength and hardness. Most steels used for structural work, bridges, ships, boilers, and pressure vessels contain between 0.8% and 1.6% manganese.
b) *Chromium* also increases strength and hardness, and makes it easier to harden the steel, i.e. it gives a slower critical cooling rate. Additions in excess of 13% confer good corrosion resistance.
c) *Nickel*, usually added with other elements, increases strength and improves the impact properties at low temperatures – a steel containing 9% nickel is used for storage tanks containing liquid methane at −162°C.
d) *Molybdenum* in amounts between 0.5% and 1.0% improves creep strength. Normally it is added together with 1.5% to 5% chromium.

Although *silicon* is occasionally used as an alloying element, it is mainly added to the molten steel ingot to reduce the oxygen content. In most steels a residual amount of silicon is present but it does not contribute to the properties.

10.6 Selecting carbon steels

Steels offer a wide variety of properties which must be matched to service or manufacturing requirements if the best grade is to be selected. The following examples have been chosen to illustrate some of the factors which are taken into account in typical applications.

Sheets for pressing into car bodies Steel sheets which are to be pressed into body panels must have good ductility and must not work-harden rapidly. The force–elongation curves should show an appreciable amount of plastic strain before fracture, and the yield point should not be high since this could call for a more powerful press. These requirements can be met if the pearlite content is low and the steel is almost entirely ferritic. The sheets must also have uniform thickness and good surface finish, which means that the steel will have been cold rolled during manufacture.

Typically, steels used for car bodies, and for similar applications, contain between 0.07% and 0.12% carbon and less than 0.5% manganese. Corresponding mechanical properties would be 210 N/mm^2 yield stress, 410 N/mm^2 tensile strength, 44% elongation, and 61% reduction of area.

Structural work Structures such as bridges, building frames, and storage tanks are generally designed on the basis of the yield stress. Quite often the maximum allowable design stress is two thirds of the yield stress, and a steel must be chosen which meets this requirement. However, the choice of a steel is frequently limited by the need to be able to form the plates or sections by bending, and to weld them together. This sets an upper limit to the carbon content, and manganese is added to bring the strength up to the desired level. British Standard BS 4360:1979 covers a range of steels which are suitable for most structural applications. These are divided into four grades giving yield stresses ranging from 230 to 450 N/mm² (Table 10.4). These steels usually contain 0.14% to 0.2% carbon and 0.8% to 1.2% manganese. The relative amounts of these two elements are adjusted by the steelmaker to give the required mechanical properties.

Table 10.4 Structural steels (BS 4360:1979)

Grade	Yield strength (N/mm^2)	Tensile strength (N/mm^2)	Elongation (%)
40	230–260	400–480	22
43	240–280	430–510	20
50	340–355	490–620	18
55	415–450	550–700	17

Note: the grades are further subdivided according to Charpy impact values, for example:

Grade	Temperature of specimen during test (°C)	Minimum Charpy test value (J)
40A	Not specified	Not specified
40B	Room temperature	27
40C	0	27
40D	−10	41
	−20	27
40E	−20	61
	−30	47
	−50	27

Frequently, in structural applications there is also a requirement for good impact properties, especially at temperatures around freezing point and below. The steels in BS 4360 are further classified according to Charpy value.

Casting Metals for casting must have acceptable fluidity. The fluidity in steels is not good but can be improved by the addition of about 0.6% silicon. If it is to be used for casting, a steel must have high resistance to cracking, which means that the carbon content should not be high. A number of carbon steels which are suitable for casting are listed in BS 3100:1976 (see Table 10.5). These have carbon contents of less than 0.45%, and increased strength is achieved by manganese additions.

Table 10.5 Carbon-steel castings for general purposes (from BS 3100:1976)

Grade	Composition (maximum %) Carbon	Manganese	Silicon	Tensile strength (N/mm^2) (minimum)	Elongation % (minimum)
A1	0.25	0.9	0.6	430	22
A2	0.35	1.0	0.6	490	18
A3	0.45	1.0	0.6	540	14
A4	0.25	1.6	0.6	540	16
A5	0.33	1.6	0.6	620	13

Note: the castings can be heat-treated to achieve the specified properties (see page 155).

Machine components A large variety of machine parts such as gear wheels, shafts, connecting rods, agricultural tools, and turbine discs are made by forging and machining. Steels for these applications must have not only adequate strength at room temperatures but also high ductility at forging temperatures (750°C to 1200°C). A further requirement may be for heat treatment to improve wear resistance or toughness. Generally steels for this type of application contain 0.35% to 0.6% carbon and are typified by BS 970:1972 (see Table 10.6).

Some steels tend to tear during machining, reducing cutting speeds and giving a poor surface finish. Better results can be obtained by using a steel which contains 0.1% to 0.2% sulphur or lead. *Free-cutting steels*, as they

Table 10.6 Mechanical properties of medium-carbon steels (from BS 970: part 1:1972)

	Steel designation '40' carbon (080 M40)	'50' carbon (080 M50)	'55' carbon (070 M55)
Carbon min. %	0.32	0.45	0.50
max. %	0.40	0.55	0.60
Manganese min. %	0.60	0.60	0.50
max. %	1.00	1.00	0.90
Yield stress (N/mm^2) (min.)	275	310	355
Tensile strength (N/mm^2) (min.)	540	620	695
Elongation %	16	14	12

Note: properties quoted are for the normalised condition (see page 151).

are called, are available in a range of carbon contents as bar or rod. Their main disadvantage is that they crack when arc welded, as they are weak at the solidification temperature. Studding which is to be welded into place and shafts which have welded flanges must be made from steels with a lower sulphur content, even though this means some sacrifice of machinability. The alternative is to use a process such as friction welding or flash welding which does not involve melting.

11 Heat treatment of steels and cast irons

The dependence of the structure of steel and cast iron on the cooling rate opens up the possibility of using heat treatment to produce a range of properties.

11.1 Response of steels to heat treatment

Three heat treatments are commonly applied to hypo-eutectoid carbon steels: annealing, normalising, and hardening. In each of these, the steel is heated to 50°C above the A_3 point (fig. 11.1). The structure is austenitic at this temperature, and all the carbon is taken into solution. The properties of the steel depend on how fast it is cooled from the austenitic condition to room temperature. Essentially, the difference between the three heat treatments listed is the rate of cooling from the A_3 point.

Full annealing is an approach to equilibrium conditions. The steel is cooled in the furnace so that the cooling rate can be controlled by

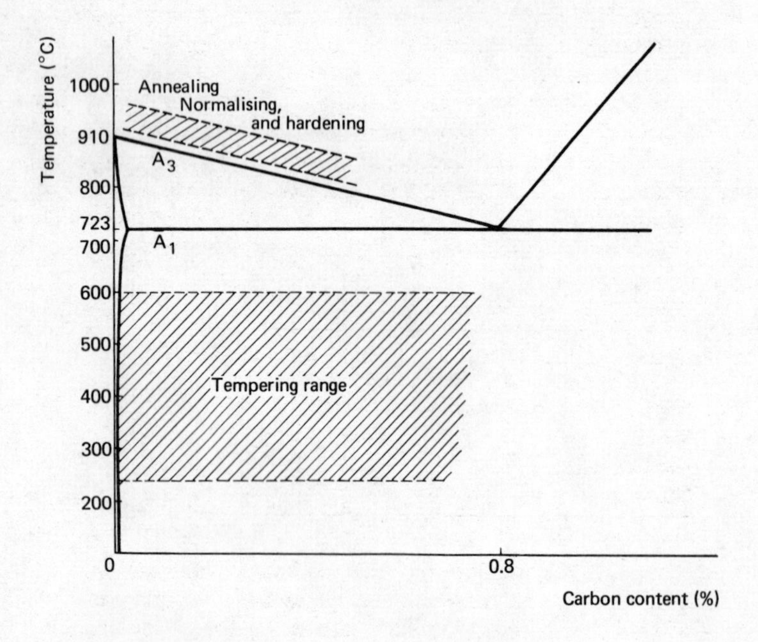

Fig. 11.1 Heat-treatment ranges for carbon steels

gradually reducing the heat supplied, thus allowing the γ-to-α transformation to go to completion at each temperature. The structure consists of large grains of ferrite with coarse pearlite in which there are thick plates of ferrite and carbide. This gives the softest possible structure and is an ideal preparation for mechanical working. Low yield stress and tensile strength are associated with annealing.

Normalising is used as a finishing treatment for carbon steels as it gives higher strength than annealing without serious loss of ductility. The steel is removed from the furnace and allowed to cool in air, so that the cooling rate is faster than in annealing. The final grain size is smaller than with annealing because a larger number of grains are nucleated and there is less time for them to grow.

At low carbon contents, say 0.2%, the effect of this small grain size is more noticeable in impact strength than in tensile strength. The latter increases by only about 5% compared with annealed material, but impact strengths (Charpy value) can be improved by as much as 20% by normalising. Steel plates are often delivered in the normalised condition when high impact-test values are specified, especially at low temperatures, e.g. BS 4360 grades 40D and 40E (see Table 10.4). At 0.4% carbon, the effect on strength also becomes marked, and improvements of 15% in the yield stress are obtainable.

Hardening relies on exceeding the cooling rate needed to produce a martensite structure (page 144). This involves removing the steel from the furnace and plunging it immediately into a quenching medium.

There is little advantage to be gained from cooling faster than the critical rate (see page 145) – on the contrary, distortion and cracking occur with very fast cooling. The quenching medium is therefore chosen to cool at a rate which just exceeds the critical value. Three commonly used quenching media are listed in Table 11.1; salt solution (brine) gives the fastest cooling rate. Brine would be used for an 0.35% carbon steel which is difficult to harden, while oil would be suitable for a eutectoid steel.

Table 11.1 Quenching media

Quenching medium	Relative cooling rate
5% salt solution (in water)	1.0
Water at 20°C	0.85
Quenching oil	0.42
Mineral oil	0.30

The effectiveness of the medium depends to some extent on the techniques used for quenching. As soon as the steel component is plunged into the quenching bath, the liquid in contact with the surface of the metal is vapourised. A layer of vapour is established over the surface which effectively insulates the steel and slows down the cooling rate. The component must be agitated to break up this film and to ensure that fresh liquid is continually brought into contact with the metal. In this way, heat is conducted rapidly from the surface and the maximum cooling rate is achieved.

Another factor which affects the cooling rate is the temperature of the quenching medium. The rate is reduced by 40% if the temperature of the water in the quenching bath is raised from 20°C to 40°C. At boiling point the rate is only 10% of that at 20°C.

11.2 Hardenability

When a bar of steel is quenched, the rate of heat removal at the surface is governed by the effectiveness of the quenching medium. On the other hand, the heat in the body of the bar must first flow to the surface where it can escape (fig. 11.2). The cooling rate inside the bar is controlled by the thermal conductivity of the steel. It also depends on the temperature gradient within the bar. The faster the surface temperature is lowered, the quicker the heat flows from the centre outwards. However, the cooling rate within the bar is always slower than at the surface. There is a possibility, therefore, that at some point below the surface the cooling rate

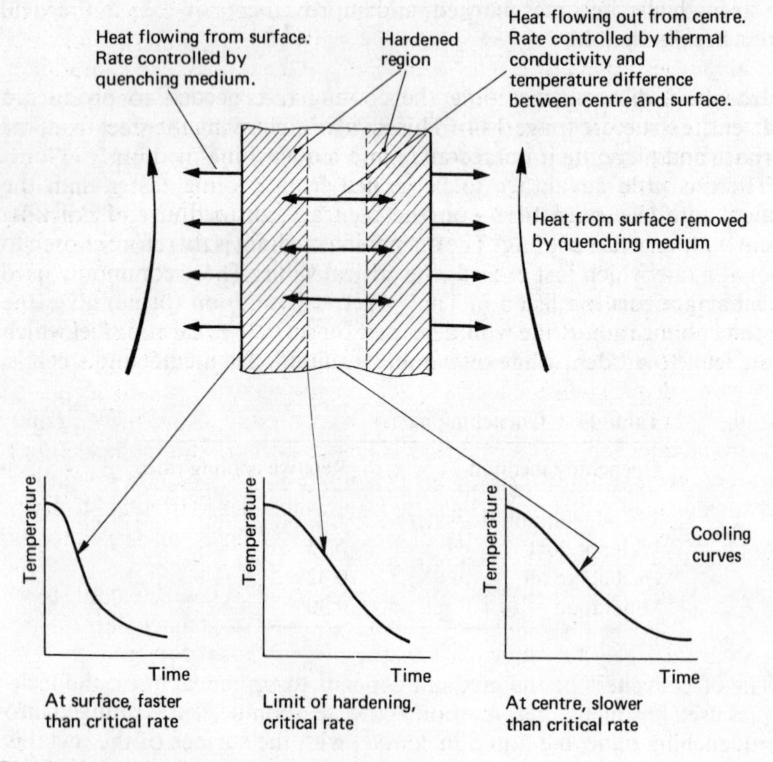

Fig. 11.2 Depth of hardening in steel

could be less than the critical value, with the result that the metal would not harden.

The hardenability of a steel is a measure of the depth of hardening when it has been quenched. If a steel has good hardenability, even thick material has uniform hardness through the complete cross-section. By contrast, a bar made from a steel with low or poor hardenability has a soft centre (or core) after quench hardening.

The hardenability of a steel depends on three prime factors:

a) the composition of the steel – steels which have a low critical cooling rate have better hardenability; hence steels with a high carbon content harden to a greater depth than low-carbon steels. Alloying elements such as chromium improve the hardenability of a steel.
b) the quenching medium.
c) the section dimensions (size of component).

The hardenability of a steel is indicated in Standards by stating the *ruling section*. This is the maximum thickness which can be used and still achieve the stated properties through the complete section. In other words, it is the maximum depth to which the steel hardens by a specified amount on quenching from above the A_3 temperature.

Size is also important in relation to distortion. Variations in the rate of cooling within the component can lend to different amounts of contraction at various points through the thickness. One part of the component can contract more than another, and the shape or dimensions can be changed. To some extent this can be minimised by the technique used to lower the part into the quenching bath. Thus with thin sections the long axis should be vertical. In this way, the heat flows quickly through the thin section with the result that the temperature distribution is relatively uniform.

11.3 Surface hardening

In some applications there is a need for the surface to be hard, to resist wear, while the body of the component remains ductile. There are two ways of hardening the surface of a steel component.

In the first method, a thin surface layer is raised to above the A_3 point and is then quenched. Either oxygen–fuel-gas burners (flame hardening) or induction coils can be used to heat the surface. The method is suitable only for steels with more than 0.4% carbon, which means that the strength of the unaffected metal in the body of the component will be relatively high.

As an alternative, the composition of the surface layers of a ductile low-carbon steel can be altered to make them responsive to quench hardening. One variant of this technique (case hardening) involves enriching the surface with carbon. Steels used for case hardening contain about 0.2% carbon, and the aim is to increase this to between 0.8% and 0.9% to a depth of 0.5 mm to 1.0 mm. To achieve this, the component must be heated in a carbon-rich atmosphere. It can be packed into a box containing charcoal, or immersed in a bath of fused sodium-cyanide/soda-ash mixture, or kept in a methane atmosphere during heating. With an 0.2% carbon steel, the component must be heated to 900°C to 950°C and held at this temperature for about three hours while carbon diffuses into the surface layers. The component is then quenched to form martensite in this carburised metal. As the core is still low in carbon, it does not harden. Sometimes a second heat treatment may be necessary to adjust the properties of the core. For example, the grain size of the low-carbon-steel core may increase during the prolonged heating at 900°C, with a consequent loss of mechanical properties. A second heating and quenching can reduce the grain size in the core.

11.4 Tempering

Steel sections which have been hardened by quenching are brittle and contain stresses which result to some extent from one part of the component cooling faster than another. The transformation from austenite to martensite also creates internal stress. These stresses can be reduced to a very low level by tempering. This consists of heating the hardened component to between 200°C and 600°C and holding it at temperature for a predetermined period of time. Tempering also alters the mechanical properties.

At the lower end of the temperature range there is very little effect on the structure but some of the stresses are relieved. Above 230°C, carbon atoms which were trapped in the lattice during quenching move from these non-equilibrium positions and form small particles of cementite. The strain in the lattice is reduced by the removal of the carbon atoms, and the hardness of the martensite is lowered. This is accompanied by an increase in the toughness and impact strength of the steel. The extent of the softening is a function of the temperature used for the tempering operation – the amount of softening is greater at higher temperatures. As the temperature is raised from 230°C, progressively more carbon is released from the lattice, the cementite particles become larger, and both the hardness and the strength fall.

The steel must be held at temperature long enough to allow all the related changes to take place. Tempering temperatures must not exceed the A_1 point, as the steel would become partially austenitic and the benefits of the hardening treatment would be lost.

In heat-treatment practice, the tempering temperature is related to the function of the component. Tools are treated within the range 230°C to 300°C. From 230°C to 260°C the aim is to give a lasting cutting edge with good abrasion resistance. Tempering between 270°C and 300°C confers better shock resistance and springiness while retaining a good cutting edge. Traditionally the temperature is judged by the colour of the oxide film which forms on the surface of the steel (Table 11.2).

Although temper colours have been widely used for many years, more accurate methods of temperature measurement are to be preferred.

Table 11.2 Tempering temperatures for tools

Temperature (°C)	Colour of oxide on surface	Types of tool
230	Pale straw	Planing tools
240	Dark straw	Milling cutters
250	Brown	Taps and dies
260	Purplish-brown	Punches; drill bits
270	Purple	Press tools
280	Dark purple	Cold chisels
300	Blue	Wood saws; springs

If greater ductility and toughness are required, for example in shafts and high-strength bolts, the steel is tempered at 300°C to 600°C.

11.5 Choice of heat treatment

The choice of the most suitable heat-treatment sequence is dictated by what is to happen to the steel. If it is to be subjected to further processing, either recrystallisation or full annealing would be appropriate. On the other hand, heat treatment during the finishing stages is often specified to ensure that properties match specification. Quite often this is covered by a Standard. The information from BS 3100:1976 in Table 11.3 illustrates the way that the manufacturer is guided in the choice of heat treatment while still being allowed some latitude where properties are easily achieved (grades A4 and A5).

Finally, the heat treatment may be aimed at the production of a wear-resistant surface or a cutting edge which does not blunt rapidly.

11.6 Heat treatment of cast iron

Both grey and white cast irons lack ductility, which limits their use. S.G. irons have provided a useful alternative to grey irons, offering good ductility and strength due to the graphite being in the form of nodules rather than flakes.

In spite of their lack of ductility, the very good fluidity of white irons makes them suitable for casting sections which would be too thin for either grey or S.G. irons. The brittleness of a white iron can be removed by heat treatment at 850°C to 1050°C, converting it to a *malleable iron*. As a result of non-equilibrium cooling, the structure of a white iron is pearlite plus cementite. The prolonged heating, up to 100 hours, decomposes the cementite, forming spheroidal-graphite aggregates or rosettes.

There are three types of malleable cast iron: whiteheart, blackheart, and pearlitic. They are produced by different heat-treatment cycles and differ in tensile properties (Table 11.4). All three varieties have acceptable ductility and are used for pipe fittings and flanges, wheel hubs, cam shafts, rocker arms, universal joints, and many other vehicle components.

Table 11.3 Properties and heat treatment of steel castings (from BS 3100:1976)

1.5% manganese steel castings for general purposes

Chemical composition The steel shall contain:

Element	A4		A5 and A6	
	% min.	% max.	% min.	% max.
Carbon	0.18	0.25	0.25	0.33
Silicon	—	0.60	—	0.60
Manganese	1.20	1.60	1.20	1.60
Phosphorus	—	0.050	—	0.050
Sulphur	—	0.050	—	0.050

Heat treatment All castings shall be supplied in the heat-treated condition, the heat treatment being in accordance with the requirements specified in (a) to (c) below.

a) *Steel A4* The final heat treatment shall consist of normalising, normalising and tempering, or hardening and tempering at suitable temperatures to give the mechanical properties specified.

b) *Steel A5* The final heat treatment shall consist of normalising, normalising and tempering, or hardening and tempering at suitable temperatures to give the mechanical properties specified for this steel, which is suitable for castings having section thicknesses of up to 100 mm.

c) *Steel A6* The final heat treatment shall consist of oil or water hardening and tempering at suitable temperatures to give the mechanical properties specified for this steel which, in this condition, is suitable only for castings having section thicknesses of up to 63 mm. If greater section thicknesses are involved and similar mechanical properties are required, it is necessary that a steel of higher hardenability, e.g. steel BT1, be specified.

Mechanical properties The mechanical properties to be obtained from test-pieces selected, prepared and tested in accordance with the requirements of section 1 of this standard shall be as follows.

Mechanical property	A4	A5*	A6†
Tensile strength			
N/mm^2, min.	540	620	690
N/mm^2, max.	690	770	850
Lower yield stress or 0.2% proof stress, N/mm^2, min.	320	370	495
Elongation, % min.	16	13	13
Charpy V-notch impact value, J, min.	30	25	25

* Limiting section thickness 100 mm
† Limiting section thickness 63 mm

Brinell hardness of castings If required by the purchaser and stated on the order, castings heat treated to the requirements of this material specification shall have a Brinell hardness within the following ranges:

steel A4 152 HB to 207 HB
steel A5 179 HB to 229 HB
steel A6 201 HB to 225 HB

Table 11.4 Typical mechanical properties of malleable irons

Type of iron	0.2% proof stress (N/mm^2)	Tensile strength (N/mm^2)	Elongation (%)	Hardness (HB)
Whiteheart	210	360	5	160
Blackheart	185	310	10	120
Pearlitic	450	570	4	230

Note: hardnesses are measured with a Brinell ball indentor.

11.7 Furnace atmosphere during heat treatment

In the preceding sections, much emphasis has been placed on the roles of temperature and cooling rate. Another very important aspect of heat-treatment practice is control of the atmosphere which surrounds the component when it is in the heat-treatment furnace.

In general, it is undesirable for the composition of the steel to be altered at the surface by either the removal or the addition of other elements. There are, of course, exceptions to this – when a steel is being carburised, the aim is to increase the carbon content; whereas the furnace atmosphere in the production of whiteheart malleable iron is purposely made oxidising to achieve the opposite effect, i.e. to reduce the level of carbon in the iron. In the main, however, the composition of the steel should be the same when the steel comes out of the furnace as it was when it went in.

Similarly, the surface condition should not deteriorate. In a furnace atmosphere which has an excess of oxygen, the surface of the steel becomes oxidised. Light uniform oxidation is often acceptable even on a finished article but can easily be removed by dipping in an acid bath if a clean surface is needed. Although heavy oxidation or scaling can be treated in the same way, there is a danger that the greater length of time in the acid bath could lead to severe pitting of the surface, which would be unacceptable for many applications. These pickling operations also create production problems – ventilation must be provided to remove the fumes given off from the bath, and there are many hazards in the handling of the acid. Furnace-atmosphere control offers a better method of producing surfaces free of heavy oxidation.

11.8 Types of heat-treatment furnace

There are three main types of furnace used in the heat treatment of steel: direct-fired, muffle, and salt bath.

Direct-fired furnaces (fig. 11.3(a)) are principally used for large items – for example, the annealing of plates during forming operations. They can be gas- or oil-fired, and the products of combustion from the burners are allowed to come into contact with the plate or component. Rates of heating are high, but it is not easy to achieve uniform temperature

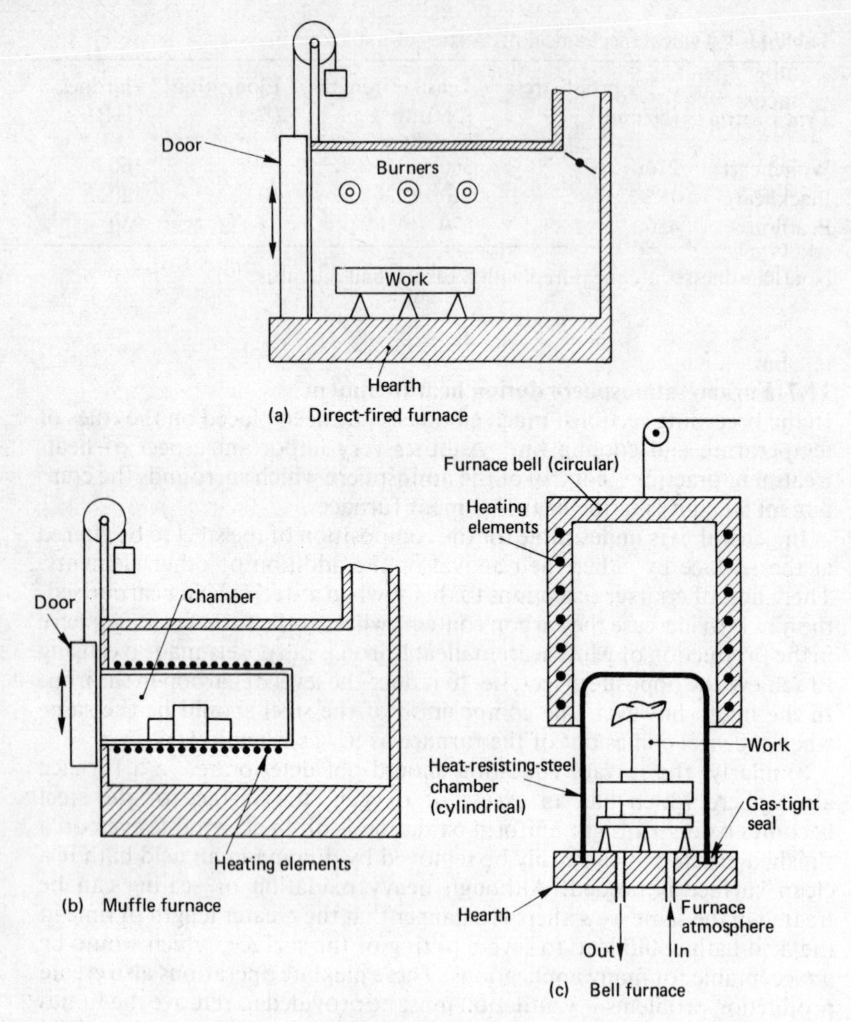

Fig. 11.3 Direct-fired, muffle, and bell heat-treatment furnace

distribution. A small amount of control can be exercised over the composition of the atmosphere in the furnace by adjusting the fuel/air ratio at the burners, but this is limited by the need to ensure efficient combustion. Indirectly heated furnaces are used where it is desirable to control the atmosphere surrounding the steel component.

Muffle furnaces In a *muffle furnace* (fig. 11.3(b)), the work is held in a chamber made of a heat-resistant material which is impervious to gases. The walls of the chamber are heated by a flame or electric elements and radiate heat to the work.

A *bell furnace* (fig. 11.3(c)) is a special type of muffle in which the chamber is made from a heat-resisting steel and stands on a hearth. The furnace bell is lowered over the chamber. When the heating cycle is completed, the bell is removed and is used to heat another chamber while the first is cooling down.

Rates of heating are slow in a muffle-type furnace, but the atmosphere in the chamber can be controlled to avoid oxidation. If needed, the atmosphere can be made reducing, to give a bright finish, or carburising. Muffle furnaces are used for small to medium-sized components.

Salt-bath furnaces (fig. 11.4) are also indirectly heated and are widely used for the heat treatment of small tools, gears, and other machine parts. The salt is melted in a heat-resisting steel pot and the component is immersed in it. The transfer of heat is very efficient because the molten salt is in intimate contact with the surface of the metal. Good temperature control can be exercised, and temperature distribution in the workpiece is uniform. This makes salt baths ideal for tempering operations.

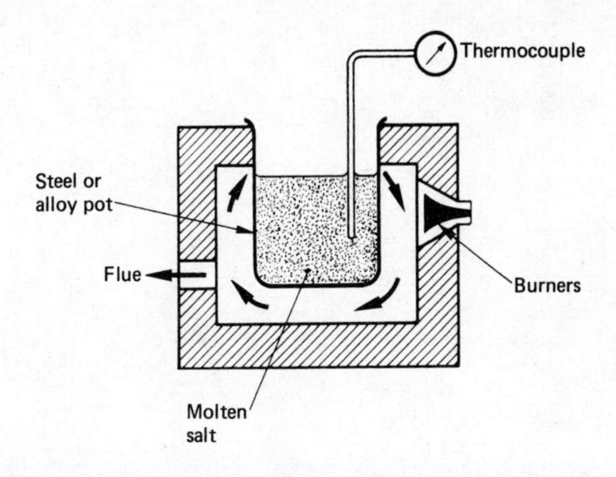

Fig. 11.4 Salt-bath furnace (gas- or oil-fired)

The operating temperature of the bath must be well above the melting point of the salt, to avoid local freezing when cold components are loaded. At the same time, the boiling point of the salt must not be exceeded. A range of salts is available to suit the operating requirements. In the main, these salts are neutral in that they do not react with the steel, but they may dissolve iron oxide to give a clean surface. On the other hand, a salt may be chosen with the specific intention of modifying the steel. As we saw in the section on surface hardening, a salt containing sodium cyanide carburises the surface of the work.

The use of salt baths must be subject to strict safety measures. All heat-treatment processes involve hazards arising from the need to handle hot metal, but salt-bath treatments pose additional problems. Care must be taken to avoid splashing when the component is inserted into the bath. The salt may be poisonous (for example, sodium cyanide) – it must be stored securely, and contact with the skin must be avoided. Effluent from washing processes designed to clean adhering salt from the components must be disposed of safely and not be allowed to contaminate the drains. Provision must also be made to remove fumes which may be given off from the bath. Given good shop planning and supervision, however, salt baths can be safe and efficient and they are widely and successfully used for the heat treatment of steel.

Index